MOLECULAR REGULATION of AROUSAL STATES

CRC Press
METHODS IN THE LIFE SCIENCES

Gerald D. Fasman - Advisory Editor
Brandeis University

Series Overview

Methods in Biochemistry
John Hershey
Department of Biological Chemistry
University of California

Cellular and Molecular Neuropharmacology
Joan M. Lakoski
Department of Pharmacology
Penn State University

Research Methods for Inbred Laboratory Mice
John P. Sundberg
The Jackson Laboratory
Bar Harbor, Maine

Methods in Neuroscience
Sidney A. Simon
Department of Neurobiology
Duke University

Joseph M. Corless
Department of Cell Biology,
Neurobiology and Ophthalmology
Duke University

Methods in Pharmacology
John H. McNeill
Professor and Dean
Faculty of Pharmaceutical Science
The University of British Columbia

Methods in Signal Transduction
Joseph Eichberg, Jr.
Department of Biochemical and Biophysical Sciences
University of Houston

Methods in Toxicology
Edward J. Massaro
Senior Research Scientist
National Health and Environmental Effects Research Laboratory
Research Triangle Park, North Carolina

CRC Press
CELLULAR AND MOLECULAR NEUROPHARMACOLOGY

Joan M. Lakoski, Advisory Editor

The *CRC Press Cellular and Molecular Neuropharmacology Series* provides the reader with state-of-the-art research methods that address the cellular and molecular mechanisms of the neuropharmacology of brain function in a clear and concise format. Topics covering all aspects of neuropharmacology are being reviewed for publication.

Published Titles

Molecular Regulation of Arousal States, Ralph Lydic

Forthcoming Titles

Neuropharmacology Methods in Epilepsy Research,
Steven L. Peterson and Timothy E. Albertson

Methods in Neuroendocrinology, Louis D. Van De Kar

MOLECULAR REGULATION of AROUSAL STATES

Edited by

Ralph Lydic, Ph.D.

The Pennsylvania State University
College of Medicine
Hershey, Pennsylvania

CRC Press

Boca Raton New York

Library of Congress Cataloging-in-Publication Data

Molecular regulation of arousal states / edited by Ralph Lydic.
p. cm. — (CRC Press methods in the life sciences. Cellular and molecular neuropharmacology)
Includes bibliographical references and index.
ISBN 0-8493-3361-X (alk. paper)
1. Molecular neurobiology—Laboratory manuals. 2. Arousal (Physiology)—Research—Laboratory manuals. 3. Sleep—Research--Laboratory manuals. I. Lydic, Ralph. II. Series.
[DNLM: 1. Arousal—physiology. 2. Sleep—physiology. 3. In Situ Hybridization. WL 103 M7185 1997]
QP356.2.M665 1997
612.8′21—dc21
DNLM/DLC
for Library of Congress 97-21353
CIP

International Standard Book Number 0-8493-3361-X
Library of Congress Card Number 97-21353
Printed in the United States of America 1 2 3 4 5 6 7 8 9 0
Printed on acid-free paper

The Editor

Ralph Lydic, Ph.D., is Director of Anesthesia and Neuroscience Research and Professor of Anesthesia and Cellular & Molecular Physiology at The Pennsylvania State University, College of Medicine, Hershey, PA. Dr. Lydic's research career has maintained a focus on the neurobiology of sleep and breathing. His 1979 Ph.D. in physiology from Texas Tech University used single-cell recording techniques to test the hypothesis that the onset of rapid eye movement (REM) sleep caused diminished discharge of pontine respiratory neurons. Postdoctoral years were spent in the Department of Physiology and Biophysics at Harvard Medical School. In 1981, Dr. Lydic joined the Laboratory of Neurophysiology at Harvard Medical School, where he served as Assistant Professor of Physiology. In 1986, Dr. Lydic moved his laboratory to the Pulmonary Division of The Pennsylvania State University's College of Medicine, where his research emphasized the neural control of breathing. In 1989, Dr. Lydic was appointed Director of the Division of Anesthesia and Neuroscience Research. Since 1991, he has served as Professor in the Department of Anesthesia and in the Department of Cellular & Molecular Physiology.

Awards and honors resulting from Dr. Lydic's research include an Upjohn Pharmaceutical Scholarship (Harvard Medical School); Neurobiology Program Scholarship (Woods Hole Marine Biological Laboratory); Neurobiology Program Scholarship (Cold Spring Harbor Laboratory, NY); National Research Service Award (Harvard Medical School); William F. Milton Award (Harvard Medical School); Mentor for Scholl Fellowship, National SIDS Foundation (Pennsylvania State University); Mentor for Parker B. Francis Fellowship (Pennsylvania State University); Visiting Scientist, NASA Division of Space Life Sciences, Johnson Space Center (1994 to 1995); Dunaway-Burnham Visiting Scholar, Dartmouth Medical School (1995); Mentor for Proctor and Gamble Award from the American Physiological Society (1996); and Mentor for Precollege Science Education Initiative, Howard Hughes Medical Institute (1996).

Dr. Lydic has served the American Physiological Society (APS) in a variety of offices including Chairman, Central Nervous System (CNS) Section; Program Advisory Committee; CNS Section Advisory Committee; Long-Range Planning

Committee; Chairman, FASEB Theme Committee: "Nervous System Function and Disorder"; Nominating Committee; Committee on Committees; and Public Affairs Committee.

Dr. Lydic's research program spans issues from the level of transmembrane cell signaling to integrative aspects of respiratory and arousal state control. His studies aim to elucidate the cellular and molecular mechanisms that cause respiratory depression during the loss of waking consciousness. These basic studies are funded by the National Heart, Lung, and Blood Institute of the National Institutes of Health because of their potential clinical relevance for disorders such as sudden infant death syndrome, adult sleep apnea, and anesthesia-induced respiratory depression.

Contributors

H. Elliott Albers, Ph.D.
Laboratory of Neuroendocrinology and Behavior
Departments of Biology and Psychology
Georgia State University
Atlanta, GA

Helen A. Baghdoyan, Ph.D.
Departments of Anesthesia and Pharmacology
The Pennsylvania State University College of Medicine
Hershey, PA

Radhika Basheer, Ph.D.
VA Medical Center and Harvard Medical School
Brockton, MA

Maja Bucan, Ph.D.
Center for Neurobiology and Behavior
Department of Psychiatry
University of Pennsylvania
Philadelphia, PA

Jerry J. Buccafusco, Ph.D.
Department of Pharmacology and Toxicology
Alzheimer's Research Center
Medical College of Georgia
and
Medical Research Service
VA Medical Center
Augusta, GA

Sophie Burlet, Doctorant
Department of Experimental Medicine
INSERM U, CNRS ERS
Claude Bernard University
Lyon, France

Raymond Cespuglio, Ph.D.
Department of Experimental Medicine
INSERM U, CNRS ERS
Claude Bernard University
Lyon, France

Zutang Chen
Department of Veterinary Comparative Anatomy, Physiology, and Pharmacology
Washington State University
Pullman, WA

Chiara Cirelli, M.D., Ph.D.
The Neurosciences Institute
San Diego, CA

Luis de Lecea, Ph.D.
Department of Molecular Biology
The Scripps Research Institute
La Jolla, CA

Charles W. Emala, M.D.
Department of Anesthesiology and Critical Care Medicine
Johns Hopkins School of Medicine
Baltimore, MD

Marek Fischer, Ph.D.
Institute of Molecular Biology
University of Zürich
Zürich, Switzerland

Mary Ann Greco, Ph.D.
VA Medical Center and Harvard Medical School
Brockton, MA

Steven Henriksen, Ph.D.
The Scripps Research Institute
La Jolla, CA

A. Urban Höglund, Ph.D.
Department of Comparative Medicine
Biomedical Center
Uppsala, Sweden

Thomas A. Houpt, Ph.D.
E.W. Bourne Behavioral Research Laboratory
Department of Psychiatry
Cornell Medical College
White Plains, NY

Kim L. Huhman, Ph.D.
Department of Psychology
Georgia State University
Atlanta, GA

Thomas S. Kilduff, Ph.D.
Center for Sleep and Circadian Neurobiology
Departments of Biological Sciences and Psychiatry and Behavioral Sciences
Stanford University
Stanford, CA

James M. Krueger, Ph.D.
Department of Veterinary Comparative Anatomy, Physiology, and Pharmacology
Washington State University
Pullman, WA

Clete A. Kushida, M.D., Ph.D.
Stanford Sleep Disorders Clinic and Research Center
Stanford, CA

Miroslaw Mackiewicz, Ph.D.
Division of Sleep and Chronobiology
Department of Psychiatry
University of Pennsylvania School of Medicine
Philadelphia, PA

Jean C. Manson, Ph.D.
Institute for Animal Health
BBSRC/MRC
Neuropathogenesis Unit
Edinburgh, Scotland

Patrick M. Nolan, Ph.D.
Center for Neurobiology and Behavior
Department of Psychiatry
University of Pennsylvania
Philadelphia, PA

Allan I. Pack, M.D., Ph.D.
Center for Sleep and Respiratory Neurobiology
Pulmonary and Critical Care Division
Department of Medicine
University of Pennsylvania School of Medicine
Philadelphia, PA

Maria Pompeiano, M.D., Ph.D.
The Neurosciences Institute
San Diego, CA

Tarja Porkka-Heiskanen, M.D., Ph.D.
Department of Psychiatry
Harvard Medical School and Brockton VA Medical Center
Brockton, MA

Mark A. Prendergast, Ph.D.
Department of Pharmacology and Toxicology
Alzheimer's Research Center
Medical College of Georgia
and
Medical Research Service
VA Medical Center
Augusta, GA

Lalini Ramanathan, Ph.D.
VA Medical Center and Harvard Medical School
Brockton, MA

Peter B. Reiner, V.M.D., Ph.D.
Kinsmen Laboratory of Neurological Research
University of British Columbia
Vancouver, BC, Canada

Priyattam J. Shiromani, Ph.D.
VA Medical Center and Harvard Medical School
Brockton, MA

Donna M. Simmons, HTL, Ph.C.
Department of Biological Sciences
University of Southern California
Los Angeles, CA

Gary Siuzdak, Ph.D.
Departments of Molecular Biology and Chemistry
The Scripps Research Institute
La Jolla, CA

Dag Stenberg, M.D., Ph.D.
Institute of Biomedicine
Department of Physiology
University of Helsinki
Helsinki, Finland

J. Gregor Sutcliffe, Ph.D.
Department of Molecular Biology
The Scripps Research Institute
La Jolla, CA

Irene M. Tobler, Ph.D.
Institute of Pharmacology
University of Zürich
Zürich, Switzerland

Giulio Tononi, M.D., Ph.D.
The Neurosciences Institute
San Diego, CA

Jussi Toppila, M.D.
Institute of Biomedicine
Department of Physiology
University of Helsinki
Helsinki, Finland

Hiroshi Usui, M.D., Ph.D.
Department of Molecular Neuropathology
Brain Research Institute
Nigata University
Nigata, Japan

Sigrid C. Veasey, M.D.
Center for Sleep and Respiratory Neurobiology
Pulmonary and Critical Care Division
Department of Medicine
University of Pennsylvania School of Medicine
Philadelphia, PA

Steven R. Vincent, Ph.D.
Kinsmen Laboratory of Neurological Research
University of British Columbia
Vancouver, BC, Canada

Julie A. Williams, Ph.D.
Kinsmen Laboratory of Neurological Research
Graduate Program in Neuroscience
University of British Columbia
Vancouver, BC, Canada

Lu C. Zhang, B.S.
Department of Pharmacology and Toxicology
Alzheimer's Research Center
Medical College of Georgia
and
Medical Research Service
VA Medical Center
Augusta, GA

R. Thomas Zoeller, Ph.D.
Neuroscience and Behavior Program
University of Massachusetts
Department of Biology
Amherst, MA

Rebecca K. Zoltoski, Ph.D.
Brock University
Department of Psychology
St. Catharines, Ontario, Canada

Preface

This book provides a collection of step-by-step protocols currently being used to study the molecular and biochemical mechanisms regulating arousal states. Most of these protocols focus on three naturally occurring brain states of vigilance.[1] These states include waking, deep sleep with consciousness obtunded, and the dreaming phase of sleep variously described as active, paradoxical, desynchronized, or rapid-eye-movement (REM) sleep. Different states of arousal are known to be actively generated by the brain, but the cellular and molecular mechanisms by which sleep and wakefulness are regulated remain incompletely understood. The absence of such basic information continues to retard the diagnosis and treatment of sleep disorders and state-dependent disruption of cardiopulmonary control.[2] This book is the first collection of research protocols seeking molecular level explanations for the generation of arousal states.

As with any project of this nature, this book resulted from the efforts of many people. For their endorsement, I thank Gerald D. Fasman, Advisory Editor for the CRC Methods in the Life Sciences series, and Joan M. Lakoski, Series Editor for Cellular and Molecular Neuropharmacology. Paul Petralia at CRC Press provided valuable guidance, and Pam Myers and Norina Frabotta provided invaluable help, ensuring a timely presentation of these protocols. Ultimately, the techniques presented here flow from the enthusiastic support of the authors, whom I thank sincerely.

This preface is the appropriate place to acknowledge a fundamental intellectual debt. Inherent in all the enclosed protocols is the conviction that the resolving power of reductionistic technologies can elucidate mechanisms underlying the integrative expression of physiology and behavior. The application of this goal to brain states of vigilance emulates the 19th-century European paradigm of Du Bois-Reymond, seeking to explain physiology in terms of chemistry and physics.[3] This mechanistic beacon was advanced by Jacques Loeb and J.B. Watson[4] in the Americas, where it has been refracted at various times to assert the purported ascendance of diverse approaches including behaviorism, electrophysiology, and cognitive neuroscience. In contrast, the protocols in this book do not endorse a canonical enthusiasm limited to molecular neurobiology. Most of the enclosed techniques are derivative, and their application to the study of vigilance states represents a new wave of methodological pluralism. These protocols celebrate the blurring of boundaries between reductionistic and integrative modes of investigation. These protocols show clearly that molecular technologies extend, rather than supplant, cellular and behavioral studies of vigilance states.

It is a tautology that direct studies of arousal states can only be conducted using human and nonhuman models exhibiting these states of arousal.[5] This fact points to the continued existence of long-standing tensions[3,4] inherent in the deployment of reductionistic technologies. Efforts to unify molecular and integrative study still evoke derision from a curmudgeon minority. It is fascinating to note that less than a week following the announcement of the successful cloning of an adult mammal, the British Ministry of Agriculture cut off all funding to Ian Wilmut's team at the Roslin Institute in Edinburgh. The benefits of combining integrative and molecular research programs, however, have been amply demonstrated, and each of the protocols in this book can be extended to the study of diverse physiological and behavioral phenomena. As noted elsewhere,[6] such extrapolations are a challenge but offer a creative springboard for bridging the gap between physiology and molecular biology.[7] Recognition by the National Institutes of Health that the gap between integrative and molecular biology can indeed be bridged[8] will encourage continued success in this exciting line of investigation. It is my hope that *The Molecular Regulation of Arousal States* will contribute to this success.

Ralph Lydic
Hershey, Pennsylvania
April 1997

References

1. Steriade, M., Awakening the brain, *Nature,* 383, 24, 1996.
2. Kryger, M.H., Roth, T., and Dement, W.C., *Principles and Practice of Sleep Medicine,* 2nd ed., W.B. Saunders, Philadelphia, 1994.
3. Brazier, M.A.B., *A History of Neurophysiology in the 19th Century,* Raven, New York, 1988.
4. Pauly, P.J., *Controlling Life. Jacques Loeb and the Engineering Ideal in Biology,* Oxford University Press, New York, 1987.
5. Lydic, R., Reticular modulation of breathing during sleep and anesthesia, *Curr. Opin. Pulm. Med.,* 2, 474, 1996.
6. Shuman, S.L., Capece, M.L., Baghdoyan, H.A., and Lydic, R., Pertussis toxin-sensitive G proteins mediate carbachol-induced REM sleep and respiratory depression, *Am. J. Physiol.,* 269, R308, 1995.
7. Norwood, V.F. and Gomez, R.A., Bridging the gap between physiology and molecular biology: new approaches to perpetual questions, *Am. J. Physiol.,* 267, R865, 1994.
8. RFA-HL-96015, Molecular biology and genetics of sleep and sleep disorders, *NIH Guide,* 25, P.T. 34, 1996.

Table of Contents

Chapter 1

Application of *In Situ* Hybridization to the Study of Rhythmic Neural Systems

H. Elliott Albers, R. Thomas Zoeller, and Kim L. Huhman

Contents

I. Introduction

A major challenge in studying neural systems that control rhythmicity is to demonstrate that the cells that are thought to control that rhythmicity exhibit rhythmic functional activity themselves. While there are a number of elegant examples in

0-8493-3361-X/98/$0.00+$.50

which electrophysiological techniques have been used to demonstrate rhythmic cellular activity in neurochemically defined neurons, this approach has not proven to be feasible for studying other rhythmic systems. One of the major limitations of this approach is that it is not always possible to record from cells long enough to demonstrate that their activity is rhythmic. This problem becomes increasingly difficult when the cycle or period length of the rhythm is long (e.g., circadian rhythms). Application of *in situ* hybridization has provided the opportunity to follow a functional measure of cellular activity (mRNA level) within neurochemically defined populations of neurons over an interval sufficient to demonstrate rhythmicity.

The major advantage of *in situ* hybridization over other approaches to measure mRNAs is that this technique can provide anatomical resolution at the cellular and subcellular level. Other techniques for measuring mRNA levels (e.g., Northern analysis or RNase Protection Assays) require that the area of interest be dissected out of the brain. As a result, the anatomical resolution of these other techniques is limited by the dissection skill of the investigator and the amount of target mRNA required to reach the sensitivity limit of the assay. The neuroanatomical resolution provided by *in situ* hybridization is particularly important for investigating populations of neurons that are functionally and anatomically heterogeneous. The development of methods to estimate relative differences in mRNA levels using *in situ* hybridization has provided the opportunity to examine whether mRNA levels change rhythmically in specific neurochemically defined groups of cells.

II. Step-By-Step Protocol

There are many different protocols that can be used for *in situ* hybridization. The protocol described below was chosen because it is reliable and relatively easy to use. This protocol works well with fresh-frozen tissue. The brains are removed as they would be for any other histological procedure, and the tissue is carefully frozen on finely ground dry ice to preserve the shape of the brain. The brains are stored at –80°C. We have found that six to eight brains per time point yield reliable results in our experiments, however the appropriate N may vary depending on a variety of factors.

The protocol described below can be used for the identification of mRNAs that are sufficiently abundant in individual cells. This includes most neuropeptides and enzymes critical for the synthesis of neurotransmitters. However, many receptors for neuropeptides, neurotransmitters, and steroids are not sufficiently abundant for measurement using this method and require special considerations for their detection.

A. Selection and Labeling of Probes to Recognize mRNA

The most common probes used for *in situ* hybridization are oligodeoxynucleotides and "riboprobes" (complementary RNA or cRNA probes). The following protocol employs oligodeoxynucleotide probes that can be synthesized with a DNA synthesizer or that can be purchased commercially. The sequence for many probes that recognize target mRNAs encoding proteins important for neurochemical signaling have already been successfully used in *in situ* hybridization experiments and can be easily found in published papers. However, since there are often significant differences in the nucleic acid sequence over short (e.g., 50 base pairs) regions of mRNAs from different species, it is important to use homologous probes. We label the oligonucleotide probes with S^{35} at the 3′ end using a terminal deoxynucleotide transferase kit (TdT kit) from Boehringer–Mannheim (Indianapolis, IN).

1. Add the following in sequence to a microcentrifuge tube: 4 µl 5X tailing buffer and 6 µl 5 m*M* $CoCl_2$ (both from TdT kit), 1 µl (5 pmol) of oligomer, DEPC-treated water to bring total volume to 20 µl, 50 pmol (approximately 5 µl depending on concentration of [^{35}S]dATP) (New England Nuclear, Wilmington, DE), and 2 µl TdT enzyme from the kit.
2. Vortex and centrifuge (quick spin to collect reagents at bottom of tube at 14,000 rpm) and then incubate for 15 min at 37°C.
3. Extraction #1: To each tube add 30 µl Tris (10 m*M*)/EDTA (1 m*M*; pH 8.0), 1 µl tRNA (Boehringer–Mannheim; make a 25mg/ml stock solution and store at –20°C), and 50 µl phenol/chloroform/isoamyl alcohol (Life Technologies, Gaithersburg, MD; 4°C). Vortex, centrifuge (3 min), and pipet the upper, aqueous phase into a new tube.
4. Extraction #2: To the new tube add 50 µl of chloroform/1% isoamyl alcohol. Vortex, centrifuge, and again transfer the aqueous phase to a new tube.
5. Precipitation: To this next tube add 5 µl 4 *M* NaCl and 165 µl 100% EtOH. Vortex and place in –80° freezer for 60 min. Remove tube and centrifuge for 15 min. Decant liquid and allow pellet to dry. Resuspend pellet in 100 µl Tris/EDTA for use in hybridization buffer. Test a sample of 1 µl for radioactivity, which should be around 500,000 to 1,000,000 cpm.

Note: *If background is too high following hybridization, a second ethanol precipitation can be added to quantitatively remove unincorporated ^{35}S-dATP. This will lower the cpm obtained following labeling. The number of tubes that are labeled with the above procedure depends on the number of slides to be hybridized. Each slide will need enough of this labeled probe to have 500,000 cpm.*

B. *Procedures for* In Situ *Hybridization*

To cut sections for hybridization, the brains are first removed from the –80°C freezer and allowed to equilibrate within the cryostat at a temperature range of –20 to –16°C. Coronal brain sections are cut at 12 µm and thaw-mounted onto chilled gelatin-coated slides (two adjacent sections per slide). After completion of sectioning, the dry slide-mounted sections are stored at –80°C until hybridization. The probe is hybridized on contiguous slides that are processed at the same time under the same conditions.

Note: *It is essential that equipment and buffers that come into contact with the sections before hybridization are nuclease-free. This requires the use of* *sterile slides and necessitates the treatment of all containers, slide racks, and water for wash solutions with diethylpyrocarbonate (DEPC, Sigma) followed by autoclaving.*[1]

On the day of hybridization, slides are removed from the freezer and allowed to warm to room temperature on fresh aluminum foil. They are then loaded into sterile racks. This allows all slides to be treated equivalently. Sterile plastic coplin jars can also be used if a small number of slides are to be hybridized.

1. **Prehybridization washes.** Brain sections mounted on slides (the number used depends on the variability of the signal as well as on the experimental design) are warmed to room temperature (23°C) and fixed by immersion into a 4% formaldehyde/phosphate-buffered saline (PBS; 0.55 *M* NaCl, 0.9 m*M* KH_2PO_4, 5.7 m*M* Na_2HPO_4) solution for 15 to 30 min, followed by a 2-min wash in 1X PBS. Sections are acetylated by a wash in 0.25% acetic anhydride (add acetic anhydride and shake well immediately before the wash) in 0.1 *M* triethylamine hydrochloride/0.09% NaCl for 10 min, followed by a 2-min wash in standard saline citrate (2X SSC, 300 m*M* NaCl/30m*M* sodium citrate). Sections are then dehydrated by rinsing with increasing ethanol concentrations (70, 80, 95, and 100%) and are delipidated by washing in chloroform for 5 min, followed by rinses with 100% ethanol and then 95% ethanol. Allow slides to air dry.
2. **Hybridization.** The hybridization buffer (20ml) contains 50% formamide, 4 ml 20X SSC, 200 µl tRNA (250 µg/ml), 1 ml single-stranded DNA (denatured, sheared salmon sperm, 500 µg/ml), 400 µl 50X Denhardt's solution (0.02% bovine serum albumin, Ficoll, and polyvinylpyrrolidone), 2g (10% w/v) dextran sulfate (MW 500,000), and 4.6 ml DEPC water. Calculate the number of mls of hybridization buffer needed (50 µl of hybridization buffer is pipetted onto each slide). To the buffer add enough of the labeled probe to allow 0.50×10^6 cpm per slide and add 50 m*M* dithiothreitol (DTT). A fresh stock solution of DTT should be made immediately prior to mixing the hybridization buffer (0.083 g DTT in 100 µl 0.01 *M* NaAcetate, pH 8.0; dilute 1:100 in hybridization buffer). Pipet buffer onto slide and cover with parafilm coverslips, making sure there are no air bubbles. Slides are laid flat in moist, covered containers and are incubated with the hybridization buffer for 16 h at 37°C.
3. **Posthybridization washes.** Following hybridization, coverslips are gently removed while each slide is submerged in 1X SSC. Slides are then washed as follows: two 30-min washes with 1X SSC (23°C), four 15-min washes in 2X SSC/50% formamide at 40°C, and two 30-min washes in 1X SSC (23°C). Slides are then rinsed in deionized water followed by a 5-min wash in 70% ethanol. After removal from the ethanol wash, slides are placed in a slide holder and allowed to thoroughly air dry.

If the hybridization signal is weak, two recent innovations can be considered to improve the hybridization signal and its quantitation. First, the use of ^{33}P-dATP (Andotek Life Sciences, Irvine, CA, cat # R0100) may increase the signal:noise ratio both by reducing background associated with the use of sulfur (^{35}S-dATP) and by increasing the specific activity of the probe. Second, the use of a phosphoimager instead of film to detect the radioactive signal may provide a more accurate quantitative estimate of differences in signal intensity among experimental groups.[2]

C. *Procedures for Autoradiography and Image Analysis*

There are two approaches most often used for autoradiographic analysis of the hybridization signal, film autoradiography and emulsion autoradiography. Since these two approaches can provide different types of information and can easily be done with the same sections, it is recommended that both approaches be employed. For film autoradiography, the slides are placed in light-sealed autoradiographic cassettes and apposed to Kodak (X-Omat™ or Bio-Max™) film until they are fully exposed. The exposure time will vary between probes and must be determined by pilot experiments. The films are developed manually using Kodak GBX™ developer (2 min) and Kodak GBX fixer (3 min), after which they are allowed to air dry.

A "semiquantitative" analysis of the film autoradiographs can be conducted using a computerized image analysis system.[3] The film is dimly illuminated with a Northern Light light box (Image Research Inc., St. Catherines, Ontario, Canada) and the image is fed, via a video camera (Dage-CCD 72), equipped with a 55-mm objective reversely mounted on a bellows system, to a computer with densitometry software. Many labs use Image software which can be downloaded from the NIH Image Home Page on the Internet. The system is calibrated so that the light level and camera sensitivity are consistent for all measurements. To control for the contribution of background to signal readings, the background density from an adjacent area of brain tissue is measured and subtracted from the reading of the area of interest to obtain a corrected density. For each image, the signal area is also measured. Integrated density measures are calculated (signal area multiplied by corrected density) for each signal imaged.

For emulsion autoradiography, slides are dipped in Kodak nuclear track emulsion (NTB-2 or NTB-3) and air dried for 16 h while lying on a flat surface. Dried slides are transferred to slide boxes containing desiccant, which are then made light-tight and stored at 4°C for the appropriate number of days (usually about twice the number of days needed for film exposure). Throughout this procedure, emulsion-coated slides are never exposed to light.

Slides are developed in darkness at 4°C in Kodak Dektol™ diluted 1:1 with distilled water (2 min), followed by a 30-sec rinse in distilled water, and are then fixed in Kodak fixer (3 min). Following a 20-min rinse in running tap water, the sections are counterstained in 0.2% toluidine blue, rinsed again in tap water, air dried, and coverslipped.

The emulsion-coated slides can be examined using bright-field and dark-field optics. Cells are considered to be labeled if a cluster of silver grains is associated with a single toluidine blue-stained nucleus when viewed under bright-field illumination. Silver grains over individual cell bodies within a standard can be manually counted under bright-field illumination (100 × oil immersion lens).

D. *Controls*

There are two sources of "nonspecific" labeling (background) of tissue following *in situ* mRNA hybridization: cross-hybridization to related sequences and probe "binding" to non- RNA components in the tissue. Cross-hybridization can be limited or eliminated by a combination of careful probe design and manipulation of hybridization and wash conditions. In contrast, probe binding to non-RNA components is governed by chemical interactions which cannot be addressed by alteration of stringency. Therefore, optimizing protocols represents a balance between conditions required for *selective* hybridization and conditions which limit probe binding to non-RNA components. Optimizing protocols requires the use of methods that allow measurement of these different sources of background labeling as described below.

1. **Northern analysis.** This provides the potential to demonstrate that your probe will hybridize to a single size-class of RNA among all those present in the target tissue, under similar hybridization and wash conditions. Because many different RNA molecules will co-migrate on an agarose gel, it is not possible to conclude that your probe hybridized to a single RNA molecule. In addition, you cannot preclude the possibility that your probe is hybridizing to other RNAs of different size but below the sensitivity of the assay. In contrast, identification of multiple bands on a Northern demonstrates that your probe is hybridizing to a family of RNA molecules and the *in situ* hybridization results would be difficult to interpret.
2. **Melting curve.** If your probe hybridizes to a single "target," and if this target is fully complementary to your probe, then it will have a single melting temperature. Therefore, you can perform the hybridization under the conditions described above and wash different slides at increasing temperatures. Plotting the density of the signal over the brain area of interest against the wash temperature will result in a stair-stepping decline in signal if your probe hybridizes to multiple targets of differing levels of identity. In contrast, there will be a single step over a very narrow temperature range (5°C) if there is a single target.
3. **Probes of different sequence.** This requires that you synthesize an oligodeoxynucleotide that is complementary to the target RNA over a region that is different from your experimental probe. This probe will have a completely different sequence but should label the same cells. Coincidence of labeling by two probes of dissimilar sequence is strong evidence for specificity of the hybridization.[4]
4. **Sense probes.** This requires that you synthesize an oligodeoxynucleotide that is fully complementary (i.e., mRNA-sense) to your experimental probe. Hybridization with this probe will allow you to determine the amount of signal arising from nonhybridization events. This approach assumes that the opposite DNA strand is not expressed, which may not always be a valid assumption for all eukaryotic genes.[5]
5. **RNase treatment.** Treatment of the tissue with RNase (100 mg/ml RNase A in 1X SSC, 1 m*M* EDTA, pH 8.0) before hybridization will eliminate signal resulting from hybridization. Thus, any remaining signal will represent the interaction of the probe with non-RNA components.

III. Interpretation and Limitations

In situ hybridization can be used to identify the anatomical distribution of specific mRNAs and to provide semiquantitative measures of the amount of mRNAs in specific populations of neurons. The data obtained from semiquantitative *in situ*

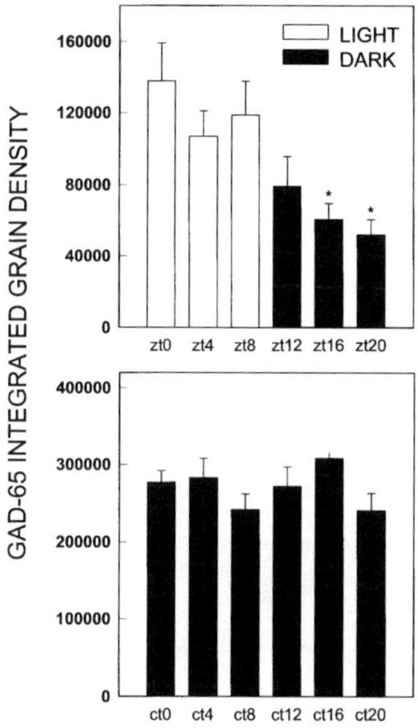

FIGURE 1.1
Example of integrated grain density (Mean ± SEM) quantified from film autoradiograms of one of the two forms of glutamic acid decarboxylase (GAD_{65}) mRNA in the rat SCN. In rats housed in a 24-hr light-dark cycle (top), GAD_{65} mRNA levels change in a 24-hr rhythm within the SCN (*less than zeitgeber time 0, $p < 0.05$). In rats housed in constant darkness for approximately two weeks, there is no statistically significant ($p > 0.05$) rhythm in the levels of GAD_{65} mRNA in the suprachiasmatic nucleus referenced to their circadian locomotor rhythm. Zeitgeber time (zt) indicates the time of day, with zt 12 indicating the time of dark onset. Circadian time (ct) indicates the phase of the circadian cycle, with ct 12 indicating the time of the onset of the circadian locomotor rhythm.

hybridization can provide extremely valuable data for investigation of rhythmic neural systems. Statistically significant differences in hybridization signal measured in the same population of cells at different phases of a rhythm can indicate whether the levels of that mRNA change rhythmically. For example, studies of the suprachiasmatic nucleus have shown that there are rhythms in the levels of several mRNAs that encode neuropeptides and enzymes important in neurotransmitter synthesis within SCN neurons[6-14] (see Figure 1.1). It has also been possible to investigate the factors that are responsible for inducing rhythmicity in these mRNAs. The 24-hr rhythms in the levels of some mRNAs in the SCN are eliminated by placing animals in constant lighting conditions, while the rhythms of other mRNAs persist.[9,12] These data suggest that the rhythms in environmental lighting induce the former rhythms, while the latter rhythms are generated by the circadian pacemaker. *In situ* hybridization has been used to study rhythms with cycle lengths ranging from minutes[15] to years.[16]

Although analysis of autoradiographs obtained with *in situ* hybridization can reliably indicate whether differences occur in mRNA levels, it does not provide a reliable indication of the absolute amount of those differences. The inability to accurately measure the absolute levels of mRNA is the result of the difficulty in preparing known quantities of target RNA *in situ* to be used as a standard curve. Therefore, if determination of the absolute levels of mRNA is critical, other techniques for measuring mRNA levels, such as RNase protection assays or quantitative PCR, should be employed.

Analysis of film autoradiographs and emulsion autoradiographs provide different, yet complementary, data on mRNA levels. Film autoradiographs are most useful for providing a semiquantitative estimate of mRNA levels within specific anatomical sites. However, it is not possible to determine whether changes in mRNA levels are the result of changes in the amount of mRNA produced by each cell or whether the number of cells producing that mRNA has changed. The cellular resolution provided by emulsion autoradiographs allows an estimation as to whether mRNA levels are changing in individual cells based on the number of silver grains over individual cells. The disadvantage of counting silver grains over cells is that it is a time-consuming and tedious process.

In our studies of circadian rhythms, we analyze both film and emulsion autoradiographs. Typically, we sacrifice groups of animals at specific intervals around the clock (e.g., every four hours) to determine whether there is a rhythm in mRNA levels. We first analyze the film autoradiographs obtained at each time point to determine whether a rhythm in mRNA levels exists. After we have established that mRNA levels vary rhythmically, we analyze the emulsion autoradiographs at the peak and valley of the rhythm (see Figure 1.2). This approach has the advantage of significantly reducing the number of cells that must be analyzed, while providing an indication as to whether the rhythm is the result of higher levels of mRNA in individual cells or whether the number of cells expressing that mRNA changes.

There are a number of limitations that should be considered when interpreting data obtained from *in situ* hybridization. Some of these limitations relate to measurements of mRNA in general, as well as to measurements by *in situ* hybridization specifically. For all mRNA measurements, it should be remembered that changes in cellular levels of mRNA may be controlled at the transcriptional or post-transcriptional level. Therefore, it may not always be accurate to discuss changes in mRNA levels as a reflection of changes in gene expression. However, changes in the cellular levels of specific mRNA will always be the result of intracellular processes governing the steady-state level of the RNA. Limitations in the interpretation of *in situ* hybridization itself, have been discussed in this chapter. Clearly, the strength of the interpretation will depend on the care with which the investigator has controlled potential confounding effects. Despite these limitations, *in situ* hybridization is a powerful technique for studying the rhythmicity of neural systems in brain.

Acknowledgments

The original research in this article was supported by NIH grants NS30022, NS34586, and NS34896.

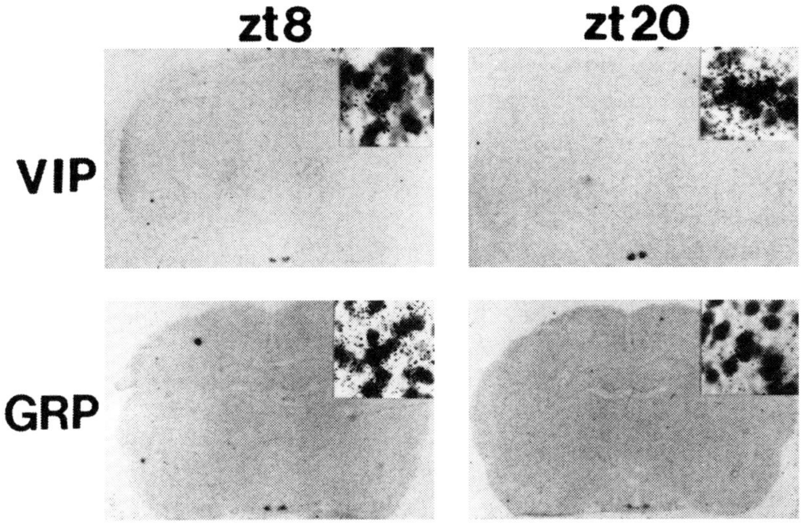

FIGURE 1.2
Autoradiographic signal following *in situ* hybridization with oligonucleotide probes complementary to vasoactive intestinal peptide (VIP) and gastrin-releasing peptide (GRP) in rats sacrificed at zeitgeber time (zt) 8 and zt 20. The hybridization signal over the suprachiasmatic nucleus on film appears as two symmetric dark areas at the base of the brain. The grain density over individual SCN neurons is illustrated with emulsion autoradiography in the inserts. The hybridization signal on film and the number of grains per cell for VIP mRNA in rats sacrificed at zt 8 is significantly ($p < 0.01$) lower than that of rats sacrificed at zt 20. In contrast, the hybridization signal on film and the number of grains per cell for GRP mRNA in rats sacrificed at zt 8 is significantly ($p < 0.01$) higher than that of rats sacrificed at zt 20.

References

1. Blumberg, D. D., Creating a ribonuclease-free environment, in *Methods in Enzymology: Guide to Molecular Cloning Techniques,* Berger S. L. and Kimmel A. R. (Eds.), Academic Press, San Diego, 1987.
2. Ito, T., Suzuki, T., Lim, D. K., Wellman, S. E., and Ho, I. K., A novel quantitative receptor autoradiography and in situ hybridization histochemistry technique using storage phosphor screen imaging, *J. Neurosci. Meth.,* 59, 265, 1995.
3. Davenport, A. P., Beresford, I. J. M., Hall, M. D., Hill, R. G., and Hughes, J., Quantitative autoradiography in neuroscience, in *Molecular Neuroanatomy,* van Leeuwen F. W., Buijs R. M., Pool C. W., and Pach O. (Eds.), Elsevier, Amsterdam, 1988, 8.
4. Azmitia, E. C., The serotonin-producing neurons of the midbrain median and dorsal raphe nuclei, in *Handbook of Psychopharmacology,* Iverson L. L., Iversen S. D., and Snyder S. H. (Eds.), Plenum Press, New York, 1978.
5. Lazar, M. A., Hodin, R. A., Darling, D. S., and Chin, W. W., A novel member of the thyroid/steroid hormone receptor family is encoded by the opposite strand of the rat c-erbA alpha transcriptional unit, *Molecular Cell Biology,* 9, 1128, 1989.

6. Stopa, E. G., Minamitani, N., Jonassen, J. A., King, J. C., Wolfe, H., Mobtaker, H., and Albers, H. E., Localization of vasoactive intestinal peptide and peptide histidine isoleucine immunoreactivity and mRNA within the rat suprachiasmatic nucleus, *Mol. Brain Res.,* 4, 319, 1988.
7. Albers, H. E., Stopa, E. G., Zoeller, R. T., Kauer, J. S., King, J. C., Fink, J. S., Mobtaker, H., and Wolfe, H., Day-night variation in prepro vasoactive intestinal peptide/peptide histidine isoleucine mRNA within the rat suprachiasmatic nucleus, *Mol. Brain Res.,* 7, 85, 1990.
8. Zoeller, R. T., Broyles, B., Earley, J., Anderson, E. R., and Albers, H. E., Cellular levels of messenger ribonucleic acids encoding vasoactive intestinal peptide and gastrin-releasing peptide in neurons of the suprachiasmatic nucleus exhibit distinct 24-hour rhythms, *J. Neuroendocrinol.,* 4, 119, 1991.
9. Uhl, G. R. and Reppert, S. M., Suprachiasmatic nucleus vasopressin messenger RNA: Circadian variation in normal and Brattleboro rats, *Science*, 232, 390, 1986.
10. Gozes, I., Shani, Y., Liu, B., and Burbach, J. P. H., Diurnal variation in vasoactive intestinal peptide messenger RNA in the suprachiasmatic nucleus of the rat, *Neurosci. Res. Comm.,* 5, 83, 1989.
11. Okamoto, S., Okamura, H., Miyake, M., Takahashi, Y., Takagi, S., Akaike, N., Fukui, K., Okamoto, H., and Ibata, Y. A., Diurnal variation of vasoactive intestinal peptide (VIP) mRNA under a daily light-dark cycle in the rat suprachiasmatic nucleus, *Histochemistry,* 95, 525, 1991.
12. Nishiwaki, T., Okamura, H., Kanemasa, K., Inatomi, T., Ibata, Y., Fukuhara, C., and Inouye, S. T., Differences of somatostatin mRNA in the rat suprachiasmatic nucleus under light-dark and constant dark conditions: an analysis by in situ hybridization, *Neurosci. Letts.,* 197, 231, 1995.
13. Okamura, H., Kawakami, F., Tamada, Y., Geffard, M., Nishiwaki, T., Ibata, Y., and Inouye, S. T., Circadian change of VIP mRNA in the rat suprachiasmatic nucleus following p-chlorophenylalanine (PCPA) treatment in constant darkness, *Mol. Brain Res.,* 29, 358, 1995.
14. Huhman, K. L., Hennessey, A. C., and Albers, H. E., Rhythms of glutamic acid decarboxylase mRNA in the suprachiasmatic nucleus, *J. Biol. Rhythms,* 11, 311, 1996.
15. Zeitler, P., Tannenbaum, G. S., Clifton, D. K., and Steiner, R. A., Ultradian oscillations in somatostatin and growth hormone-releasing hormone mRNAs in the brains of adult male rats, *Proc. Natl. Acad. Sci. USA,* 88, 8920, 1991.
16. Duncan, M. J., Cheng, X., and Heller, K. F., Photoperiodic exposure and time of day modulate the expression of arginine-vasopressin mRNA and vasoactive intestinal peptide mRNA in the suprachiasmatic nuclei of Siberian hamsters, *Mol. Brain Res.,* 32, 181, 1995.

Chapter 2

Estimation of the mRNAs Encoding the Cholinergic Muscarinic Receptor and Acetylcholine Vesicular Transport Proteins Involved in Central Cardiovascular Regulation

Jerry J. Buccafusco, Lu C. Zhang, and Mark A. Prendergast

Contents

0-8493-3361-X/98/$0.00+$.50

I. Introduction

Complex disorders such as essential hypertension are often modeled in animals subjected to some pharmacologic or surgical manipulation. However, for most cases of essential hypertension a specific causative factor is often not apparent. As with many systemic disorders, essential hypertension may have a genetic etiology in that predisposing factors can be inherited. The spontaneously hypertensive rat (SHR), one of a number of genetically induced models of hypertension, has been used widely for almost three decades as a model for the development of a significant number of cardiovascular drugs now used clinically. Estimates have ranged from a single gene with simple additive allelic effects to as few as six major genes that determine high blood pressure in SHR.[1] Studies from this and several other laboratories over the past 20 years have suggested that one inherited factor that predisposes or contributes to the development and maintenance of hypertension in the SHR (and possibly in humans) is an overactive central and perhaps spinal cholinergic (muscarinic) nervous system.[2] Drugs that enhance the synaptic levels of brain acetylcholine or directly stimulate cholinergic muscarinic receptors evoke a hypertensive response in several animal species, and in humans. The pressor response to central cholinergic receptor stimulation is even more intense and prolonged in the SHR. Conversely, depletion of neuronal acetylcholine or blockade of central muscarinic receptors in certain brain regions produces a profound fall in blood pressure in the SHR. Again, the latter response is quite modest in normotensive controls.

The ability of cholinomimetic drugs to evoke exaggerated hypertensive responses in SHR is suggestive of an alteration in the function of cholinergic neurons or their receptive sites. While alterations in the brain acetylcholine synthetic and degradative enzymes purported to exist in hypertensive compared with normotensive rats suggest some derangement in cholinergic neurobiology in SHR,[2] enzyme markers are not always good predictors of cholinergic neuronal function.

Five muscarinic receptor genes (M1–M5) have been cloned which encode distinct muscarinic cholinergic receptors.[3] Gene products for M1, M3, and M5 receptors correspond to respective receptors that activate phospholipase C via a pertussis toxin-insensitive G-protein. These subtypes are thought to mediate primarily excitatory synaptic transmission. M2 and M4 receptors inhibit adenylate cyclase activity via a pertussis toxin-sensitive G-protein and mediate primarily inhibitory synaptic transmission. Because of their structural homology and pharmacological similarity, the pharmacologic ligands presently available do not clearly distinguish the five subtypes. However, estimation of mRNA levels in brain have provided receptor distribution data which is consistent with most ligand binding studies.[3] Because of their relatively low abundance in CNS tissue, the mRNA for muscarinic receptors in brain tissue has been difficult to detect and quantify by employing conventional Northern blotting techniques. In order to overcome the problem of sensitivity, we employed the reverse transcriptase–polymerase chain reaction (RT–PCR) methodology to detect the mRNAs encoding the five muscarinic receptor subtypes and the vesicular acetylcholine transporter. Quantification of mRNA does not provide direct information regarding the expression of receptor protein. However, in the case of muscarinic receptors, a wide variety of chemical stimuli and disease states known to affect the cholinergic system produce changes in cholinergic function which are reflected at the level of transcription of the receptor genes.[4-11]

In most cases, relative levels of mRNAs are sufficient, as for example in determining whether levels of M2 receptor mRNA is different between SHR and normotensive controls. There exist many excellent approaches to the relative quantification of mRNA in tissue samples as measured by RT–PCR. Several such methods use internal cRNA or cDNA standards that are either coamplified (coreverse transcribed in the case of a cRNA) or competitively amplified to control for intersample variability. Such methods can be limited by the complexity of the assay, specialized assay conditions for coamplification, and lack of adequate controls for starting mRNA quantity and quality.[12] In our studies, most often we have used the amplification of the endogenous internal standard glyceraldehyde 3–phosphate dehydrogenase (G3PDH), that is a ubiquitously expressed housekeeping gene in a parallel tube to control for baseline mRNA quantity and quality. Although we include no control for inter-sample variability, this issue can be addressed by strict standardization of all assay conditions and the demonstration of reproducibility. Thus, we routinely employ several measures to ensure that the PCR products amplified using the RT–PCR method are valid and reproducible:

1. The relationship between the amount of product detected and the cycle number should be log-linear for target and internal standard amplicons. It is also helpful (but not necessary) that the slopes of both curves are similar, indicating similar amplification efficiencies.
2. Total RNA should be efficiently extracted from tissue samples measured spectrophotometrically to provide a uniformly constant amount of baseline RNA for each sample prior to amplification.
3. Standard samples should be routinely re-assayed and in different wells of the thermal cycler to calculate the coefficient of variation — which should remain below 10%.
4. An amplification cycle number should be used that is on the exponential phase for each amplicon.
5. The linear range of the assay should be determined either by amplifying varying concentrations of a standard RNA extract, or by running new samples at three or more PCR cycles to determine if the amplification efficiency remains constant.
6. Data should be expressed as the ratio of the amount of cDNA product (target gene)/cDNA product for the standard derived from the same sample of RNA.

II. Extraction of Total RNA from CNS Tissue for RT–PCR

The relative instability of single-stranded mRNA in both *in vivo* and *in vitro* preparations, relative to DNA, significantly complicates the daily use of this technique.[13,14] Thus, this initial step in conducting the RT–PCR reaction provides the most potential for experimental error, and critical aspects of method preparation must be addressed prior to cDNA amplification.

1. **Inactivation/exclusion of exogenous RNases**. Significant RNase activity may be introduced by various sources including experimenter clothing, hands, airborne dust, and laboratory materials used during RNA isolation. The following steps may be taken to minimize exposure to these substances:
   a. Dedicated clothing (such as laboratory coats, surgical scrubs) should be maintained.
   b. Surgical gloves should be worn at all times during tissue preparation and RNA isolation, and they should be changed frequently.
   c. All glassware, pipets, pipet tips, etc., should be dedicated to RNA methods and should be treated with 0.1% diethyl pyrocarbonate in ddH_2O to inhibit exogenous RNases and subsequently autoclaved.
   d. Overnight flooding of the work space to be used in RNA isolation by UV (254 nm) illumination to degrade exogenous RNases and sources of contaminating DNA.
2. **Inactivation of endogenous RNases.** Endogenous RNase activity in tissues represents an unavoidable source of total RNA degradation which, if not controlled for, can lead to significant loss of total RNA content prior to completion of the isolation process. Inclusion of chaotropic agents such as guanidinium in extraction buffers provides significant inhibition of endogenous RNase. A typical starting concentration in self-initiated isolation protocols is 4 *M* guanidine thiocyanate, however, concentrations can be optimized for extraction from different tissue sources.[15] Addition of β-mercaptoethanol is also recommended because it provides additional inhibition of RNase.[15] Commercially available RNA isolation kits, such as that described below, typically include proprietary mixtures of detergents, chaotropic agents, salts, etc., optimized to inhibit RNase in tissues.

A. *Tissue Collection*

For collection of all CNS tissues, animals are sacrificed by rapid decapitation and tissues immediately extracted from the cranium or vertebral column over ice, frozen in liquid nitrogen, and stored at –70°C until assay. Non-degraded total RNA is isolated using the FastRNA™ Kit-Green (Bio 101, Vista, CA) and a FastPrep™ (FP120) (Savant Instruments, Farmingdale, NY) tube shaker.

B. *FastRNA™ Protocol*

1. Add 30 to 250 mg of tissue to a 2.0 ml FastRNA tube containing RNase-free silica-ceramic beads. Add the following proprietary reagents in the order and volume that follows:

   500 μl of CRSR (chaotropic RNA stabilizing reagent)
   500 μl of PAR (acid phenol, pH 4)
   100 μl of CIA (chloroform:isoamyl alcohol, 24:1)

Note: *For tissue weights of 30 to 60 mg, it is recommended that CRSR and PAR volumes be reduced to 400 μl and CIA to 75 μl. We have found that the* *total RNA isolated may be insufficient for subsequent RT–PCR assays using larger reagent volumes.*

Caution: *Phenol is a toxin by dermal and aerosol routes. Gloves, goggles, and masks should be worn when working with phenol.*

2. Tubes are processed in the FastPrep instrument for 20 sec at speed rating 6. Following cell disruption, place the tubes on ice for 10 min because heat and positive pressure are increased within tubes during shaking.

Note: *The ceramic beads are used to provide a rapid means of cell lysis which significantly limits RNA exposure to endogenous RNases. In addition, lysis is typically more complete using this protocol than when standard glass Teflon® homogenization is used.*

3. Centrifuge the sample tubes for 15 min at 14,400 rpm at 4°C to separate aqueous phase with RNA from DNA, protein, and other cellular debris. Remove 300 to 400 μl of aqueous supernatant without disturbing the interphase.

Note: *20 to 50 μl of supernatant should be left in tube to avoid possible removal of contaminating DNA from the lower phase.*

Transfer the supernatant to a 1.5 ml sterile polypropylene tube and add 500 μl of CIA. Vortex each tube for 10 sec and centrifuge at 14,400 rpm at 4°C for 2 min to separate RNA in the aqueous phase from detergents, DNA, and protein in the lower phase.

4. Transfer 160 to 200 μl of aqueous supernatant to a sterile polypropylene tube, without disturbing the interphase. Add 500 μl of DIPS (diethylpyrocarbonate (DEPC)-treated isopropanol) to precipitate RNA. Gently mix each sample and incubate at room temperature for 2 min. Centrifuge at 14,400 RPM at 4°C for 5 min to pellet precipitated RNA.
5. Wash pellets with 250 μl SEWS (salt/ethanol wash solution, RNase-free) and centrifuge at 14,400 rpm for 1 min. Remove the liquid immediately and allow the pellets to air dry (about 10 min). Repeat this step.

Note: *SEWS should be removed with a small-bore pipet tip. Continued suspension of the pellet in remaining SEWS can lead to significant total RNA degradation.*

6. Dissolve the total RNA pellets in 100 μl of DEPC-treated ddH_2O.
7. Each sample should be assayed spectrophotometrically at 260 and 280 nm to determine nucleic acid concentration and the relative clearance of other cellular constituents. A 260/280 absorbance ratio of 1.80 or greater is sufficient to assure adequate clearance of cellular constituents other than RNA for subsequent RT–PCR reactions. A 260 nm absorbance corresponding to a total RNA concentration of 0.50 μg/μl or greater is optimal for use of most proprietary RT–PCR kits.

Note: *If starting tissue weights of approximately ≤80 mg are used, treatment of samples with DNases to degrade potentially contaminating genomic DNA may* *be unnecessary. We have found that extraction of genomic DNA by detergents is sufficient to eliminate the possibility of DNA contamination in subsequent RT–PCR reactions, as demonstrated by negative findings in samples for which RT–PCR was conducted in the absence of reverse transcription.*[16] *In many instances, however, enzymatic degradation of genomic DNA is desirable to assure the absence of possible contaminating nucleotide sequences.*

C. DNase Treatment of Total RNA Samples

DNase I (Gibco BRL, Life Technologies Inc., Grand Island, NY) may be used to treat samples of isolated total RNA using the following protocol:

1. Add the following to an RNase-free microcentrifuge tube on ice:
   1 μg RNA sample
   1 μl 10X DNase I Reaction Buffer (200 m*M* Tris HCL, pH 8.3, 500 m*M* KCL, 25 m*M* $MgCl_2$)
   1 μl DNase I, Amp Grade, 1U/μl
   DEPC-treated water to 10 μl
2. Incubate each tube for 15 min at room temperature. Inactivate the DNase I by the addition of 1 μl of 25 m*M* EDTA solution to the reaction mixture and heat samples for 10 min at 65°C.

Note: *It is important not to exceed the 15 min incubation time. Higher temperatures and longer times could lead to Mg^+-dependent hydrolysis of the RNA. EDTA must be added prior to heat inactivation to avoid this problem.*

III. RT–PCR

A. Reverse Transcription (RT)

The GeneAmp RNA PCR Kit™ (Perkin Elmer Cetus, Norwalk, CT) may be used in conjunction with a Perkin Elmer Cetus (Foster City, CA) thermal cycler (Model 480). The RT reaction procedure, as described below, is provided by Perkin Elmer Cetus. Add the reagents listed below *in the order shown* to a 500 μl polypropylene microtube.

Reagents Needed

Component/Stock Conc.	*Volume (μl/tube)*	*Final Concentration*
$MgCl_2$ solution — 25 m*M*	4	5 m*M*
10X PCR Buffer II (500 m*M* KCL, 100 m*M* Tris-HCl)	2	1X

Component/Stock Conc.	***Volume (μl/tube)***	***Final Concentration***
ddH_2O	1	
dGTP — 10 m*M*	2	1 m*M*
dATP — 10 m*M*	2	1 m*M*
dCTP — 10 m*M*	2	1 m*M*
dTTP — 10 m*M*	2	1 m*M*
RNase Inhibitor — 20 U/μl	1	1 U/μl
Reverse Transcriptase — 50 U/μl	1	2.5 U/μl
Random Hexamers — 50 μ*M*	1	2.5 μ*M*
Total RNA sample — 0.5 μg/μl	2 μl	1 μg
or		
Positive Control RNA	2 μl	10^4 copies

Note: *The 5 mM concentration of $MgCl_2$ is a suggested starting concentration. However, the optimal concentration of $MgCl_2$ for each set of primers may be determined empirically by testing concentrations from 0.5 to 5 mM in 0.5 mM increments. If samples contain chelators, or if concentrations of RNA and/or dNTPs are changed, the $MgCl_2$ concentration in the reaction mixture should be increased/decreased proportionately. Priming for cDNA synthesis can also be accomplished using Oligo d(T) or a lower (downstream) primer provided by the user. Details and indications for use of these in cDNA priming are contained in the Perkin Elmer Cetus RNA-PCR Kit.*

To reduce evaporation during thermal cycling, add 80 μl mineral oil to each microtube. Allow 10 min of incubation at room temperature for each sample to allow the extension of the hexameric primers by reverse transcriptase. Cycle the samples once at 42°C for 15 min, 99°C for 5 min, and 5°C for 5 min.

B. *PCR Amplification*

1. **Oligonucleotides.** All oligonucleotide primers were designed using the Oligo program (National Bioscience, Plymouth, MN). Those used to amplify the muscarinic mRNAs[17] were synthesized on-site using an Applied Biosystems (Foster City, CA) Model 392 nucleotide synthesizer. Oligonucleotide primers used to amplify the VAChT mRNA were purchased from the Program for Critical Technologies in Molecular Medicine (Yale University, New Haven, CT).

The following oligonucleotides were used to produce RT–PCR products for VAChT and M1–M5 muscarinic receptor subtype mRNAs:

2. **VAChT.** Upper primer: 5′-CTG GTG CTG GTC ATC GTG TG-3′ (base position 958); Lower primer: 5′-GCG AAG AGC GTG GCA TAG TC-3′ (base position 1,378). Product size = 440 bp. Genebank sequence # 507745.
3. **Muscarinic receptor subtype mRNAs.**

   M1 — Upper primer: 5′-GCA CAG GCA CCC ACC AAG CAG-3′ (base position 1,073); Lower primer: 5′-AGA GCA GCA GCA GGC GGA ACG-3′ (base position 1,425). Product size = 373 bp. Genebank sequence # M 16406.

   M2 — Upper primer: 5′-CAC GAA ACC TCT GAC CTA CCC-3′ (base position 826); Lower primer: 5′-TCT GAC CCG ACG ACC CAA CTA-3′ (base position 1,488). Product size = 686 bp. Genebank sequence # J 03025.

   M3 — Upper primer: 5′-GTC TGG CTT GGG TCA TCT CCT-3′ (base position 606); Lower primer: 5′-GCT GCT GCT GTG GTC TTG GTC-3′ (base position 1,019). Product size = 434 bp. Genebank sequence # M 16407.

   M4 — Upper primer: 5′-TGG GTC TTG GCC TTT GTG CTC-3′ (base position 461); Lower primer: 5′-TTC ATT GCC TGT CTG CTT TGT TA-3′ (base position 1,026). Product size = 588 bp. Genebank sequence # M 16409.

   M5 — Upper primer: 5′-CTG GTC TCC TTC ATC CTC TGG-3′ (base position 1,436); Lower primer: 5′-CCT GGG TTG TCT TTC CTG TTG-3′ (base position 1,809). Product size = 394 bp. Genebank sequence # M 22926.

To each RT–PCR tube add the following reagents *in the concentrations and order shown* below. To avoid variations in components between tubes, prepare a master mix of components.

Reagents Needed

Component/Stock	*Volume*	*Final Concentration*
$MgCl_2$ solution — 25 m*M*	4 µl	2 m*M*
10X PCR Buffer II — (500 m*M* KCL, 100 m*M* Tris-HCl)	8 µl	1X
ddH_2O	65.5 µl	
AmpliTaq® DNA Polymerase	0.5 µl	2.5 U/100 µl
Lower primer*	1 µl	0.15 µ*M*
Upper primer	1 µl	0.15 µ*M*

* If lower primer was used for cDNA synthesis, do not add it again. Substitute 1 µl ddH_2O.

Spin tubes for 30 to 45 sec in a microcentrifuge. For PCR amplification of positive control RNA (provided in kit), the following cycle steps are recommended.

Step	**Programs**
1	2 min at 95°C for 1 cycle
2	1 min at 95°C and 1 min at 60°C for 35 cycles

Step	Programs
3	7 min at 60°C for 1 cycle
4	10 min at 4°C for 1 cycle

For user-supplied primers, optimal conditions may be empirically determined by varying the annealing and extending temperature (37° to 65°C) and cycle number. A detailed description of optimization techniques is included with each Perkin Elmer Cetus RNA–PCR Kit.

IV. RT–PCR Product Confirmation and Quantification

Electrophoresis. To confirm molecular weights of RT–PCR products, 10 to 20 μl aliquots of cDNA samples are loaded onto a 1 to 1.8% agarose (type I: low EEO; Sigma Chemical Co., St. Louis, MO) electrophoresis gel. Dissolve the agarose in a 1X TAE solution (0.2 *M* Tris-Acetate and 4 m*M* EDTA), which also serves as the running buffer during electrophoresis. In each gel, one lane should contain a molecular weight standard (0.5 μg/lane, 100 bp DNA ladder, from Gibco BRL). Each gel is stained with ethidium bromide, and the reaction products are visualized with fluorescent illumination (254 nm). Samples are electrophoresed for 1 h at 75 V. Longer electrophoresis times and/or higher concentrations of agarose may be employed to provide greater separation of products of similar molecular weights.

Densitometry. The relative amounts of each amplicon present in the gel lanes can be estimated using standard methods of scanning densitometry. We used the IS-1000 scanning densitometer from Alpha Innotech Corp. (San Leandro, CA), but there are several instruments on the market that can adequately perform this technique. However, it is incumbent upon the investigator to establish the identity of each amplicon beyond its electrophoretic mobility. In our laboratory we have used a high performance liquid chromatographic method to identify and quantify UV peaks of interest.[18,19] We have also used selective restriction enzymes to cleave each amplicon at one or two sites, and rerun the fragments on electrophoretic gels. The new fragments should appear at expected molecular sizes on the gel. Alternatively, each amplicon can be cut and extracted from gels and submitted for sequence analysis. This approach has recently become cost effective for oligonucleotides of less than 600 bp.

V. Brain Muscarinic Receptor mRNA Expression in SHR

The specificity of the five PCR oligonucleotide primers and the identity of the respective cDNA products were determined by gel electrophoresis. A typical gel electrophoresis pattern for the PCR products obtained for all five subtypes of the

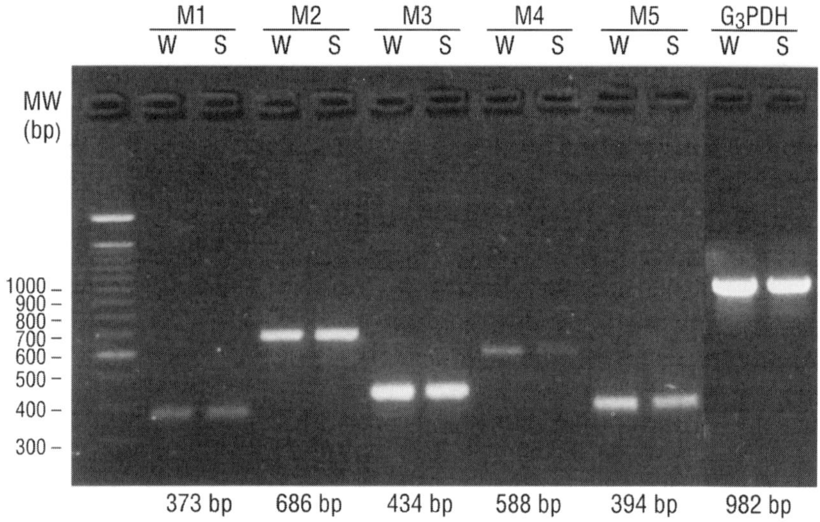

FIGURE 2.1
Agarose gel electrophoresis with ethidium bromide staining of PCR products amplified from the hypothalamus of spontaneously hypertensive rats (columns S) and from normotensive Wistar Kyoto controls (columns W) with primers specific for five muscarinic receptor genes, M1–M5, and the internal control gene, G3PDH. A DNA standard lane is shown at the left of the gel, with bands labeled in base pairs (bp). (From Wei, J., et al., *Circ. Res.*, 76, 142, 1995. With permission. © American Heart Association.)

muscarinic receptor derived from the hypothalamus of adult SHR and normotensive WKY (Wistar Kyoto) controls is depicted in Figure 2.1. All PCR products migrated in the gel according to their expected molecular weights. No amplified products were present in gel lanes where reverse transcription was omitted (data not shown). Only one band is present for each subtype, including for the control gene G3PDH.

In a separate experimental series, we attempted to determine whether changes in muscarinic receptor gene expression preceded the development of hypertension in SHR. In these experiments we used four-week-old SHR whose resting systolic blood pressure was not different from age-matched WKY. As indicated in Figure 2.2, there was a selective increase in the expression of the M2 subtype of muscarinic receptor in the SHR compared with WKY. This increased expression was also found for adult hypertensive SHR (data not shown). This finding is particularly germane to the issue of hypertension, since (as discussed in Section I.) the pressor response to stimulation of central muscarinic receptors is mediated by non-M1, possibly the M2 subtype.[2] The observation that the increased expression of the M2 receptor precedes the development of hypertension is consistent with the finding that the hypertensive response to central muscarinic receptor stimulation is exaggerated, even in these young SHR.[2] Also, it suggests that alterations in the expression of central muscarinic receptors may play a role in the initiation, as well as the maintenance of genetically induced hypertension. A detailed description of the brain distribution of the VAChT mRNA can be found in Prendergast et al.[16]

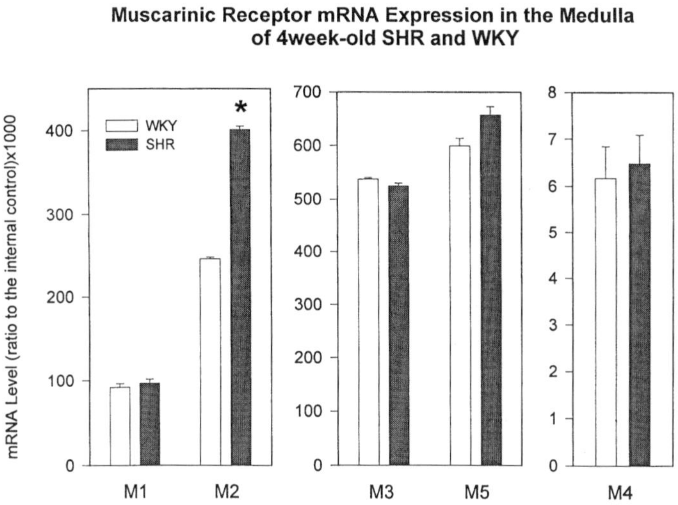

FIGURE 2.2
The level of mRNA encoding five cholinergic muscarinic receptors, M1–M5, as amplified from the medulla of four-week-old SHR and WKY rats with PCR primers specific for M1–M5 genes. Each experiment (tissue sample) was performed as duplicate PCR runs in which each amplified sample was analyzed in duplicate by HPLC. The data are expressed as the peak area ratio of subtype product/G3PDH product (internal control). Each point represents the mean ± SEM from four to six experiments.

References

1. Kurtz, T. W., Casto, R., Simonet, L. and Printz, M. P., Biometric genetic analysis of blood pressure in the spontaneously hypertensive rat, *Hypertension* 16, 718, 1990.
2. Buccafusco, J. J., The role of central cholinergic neurons in the regulation of blood pressure and in experimental hypertension, *Pharmacol. Rev.,* 48, 179, 1996.
3. McKinney, M., Muscarinic receptor subtype-specific coupling to second messengers in neuronal systems, *Prog. Brain Res.*, 98, 333, 1993.
4. Wang, S-Z., Sheng, S-Z., Joseph, J. A., and El-Fakahany, E. E., Comparison of the level of mRNA encoding ml and m2 muscarinic receptors in brains of young and aged rats, *Neuroscience Letters,* 145, 149, 1992.
5. Eva, C., Fusco, M., Bono, C., Tria, M. A., Gamalero, S. R., Leon, A. and Genazzani E., Nerve growth factor modulates the expression of muscarinic cholinergic receptor messenger RNA in telencephalic neuronal cultures from newborn rat brain, *Molec. Brain Res.,* 14, 344, 1992.
6. Asanuma, M., Ogawa, N., Haba, K., Hirata, H. and Mori, A., Effects of chronic catecholamine depletions on muscarinic M1-receptor and its mRNA in rat brain, *J. Neurol. Sci.,* 110, 205, 1992.
7. Ogawa, N., Asanuma, M., Mizukawa, K., Hirata, H., Chou, H. and Mori, A., Post-ischemic administration of bifemelane hydrochloride prohibits ischemia-induced depletion of the muscarinic M_1-receptor and its mRNA in the gerbil hippocampus, *Brain Res.*, 591, 171, 1992.

8. Wang, S-Z., Zhu, S-Z., Mash, D. C. and El-Fakahany, E. E., Comparison of the concentration of messenger RNA encoding four muscarinic receptor subtypes in control and Alzheimer brains, *Molec. Brain Res.,* 16, 64, 1992.
9. Wall, S. J., Yasuda, R. P., Li, M., Ciesla, W. and Wolfe, B. B., Differential regulation of subtypes m1-m5 of muscarinic receptors in forebrain by chronic atropine administration, *J. Pharmacol. Exp. Ther.,* 262, 584, 1992.
10. Longone, P., Moccheti, I., Riva, M. A. and Wojcik, W. J., Characterization of a decrease in muscarinic m_2 mRNA in cerebellar granule cells by carbachol. *J. Pharmacol. Exp. Ther.*, 265, 441, 1993.
11. Lee, N. H. and Fraser, C. M., Post-transcriptional regulation of the m1 muscarinic acetylcholine receptor. *Life Sci.,* 52, 562, 1993.
12. Ferré, F., Marchese, A., Pezzoli, P., Griffin, S., Buxton, E., and Boyer, V., Quantitative PCR: An overview, in *The Polymerase Chain Reaction*, Mullis, K. B., Ferré, F., and Gibbs, R.A., Eds., Birkhauser, Boston, 1994, chap. 6.
13. Sambrook, J., Fritsch, E. F., and Maniatis, T., *Molecular Cloning, A Laboratory Manual, 2nd ed.*, Cold Spring Harbor Press, Cold Spring Harbor, NY,1989.
14. Tesniere, C., and Vayda, M. E., Method for the isolation of high-quality RNA from grape berry tissues without contaminating tannins or carbohydrates, *Plant Mol. Biol. Rpts.*, 9, 242, 1991.
15. Kaufman, P. B., Wu, W., Kim, D., and Cseke, L. J., *Molecular and Cellular Methods in Biology and Medicine.* CRC Press, Boca Raton, FL, 1995.
16. Prendergast, M. A., Gattu, M., Zhang, L. C., Buccafusco, C. J., and Buccafusco, J. J., Identification of vesicular acetylcholine transporter mRNA in selected brain and peripheral tissues by RT–PCR, *Alzheimers Research,* 2, 211, 1996.
17. Bonner, T. L., Buckley, N. J., Young, A. C., and Brann, M.R., Identification of a family of muscarinic acetylcholine receptor genes, *Science* 237, 527, 1987.
18. Wei, J., Walton, E. A., Milici, A. and Buccafusco, J. J.: m1-m5 Muscarinic receptor distribution in rat CNS by RT–PCR and HPLC, *J. Neurochem.*, 63, 815, 1994.
19. Wei, J., Milici, A. and Buccafusco, J.J., Alterations in the expression of the genes encoding specific muscarinic receptor subtypes in the hypothalamus of spontaneously hypertensive rats, *Circ. Res.*, 76, 142, 1995.

Chapter 3

Voltammetric Detection of Nitric Oxide (NO) in the Rat Brain: Release Throughout the Sleep–Wake Cycle

Sophie Burlet and Raymond Cespuglio

Contents

0-8493-3361-X/98/$0.00+$.50

I. Introduction

Nitric oxide (NO), first considered a toxic gas, is now accepted as an essential biological messenger involved in several functions, e.g., immune processes, endothelium relaxation, neurotransmission, etc.[1] This diatomic low molecular weight compound is synthesized from L-arginine by NO synthases (NOS) and is produced stoichiometrically with L-citrulline. In the biological media, through which it diffuses easily, it reacts with oxygen to produce inactive metabolites like nitrites and nitrates. Within neuronal elements, NO is not stored in vesicles but is only solubilized in the cytoplasm.[2] Without storage processes and catabolizing enzymes, its production is directly ensured, depending on the need, by a family of NOS that can be divided into three essential isoforms: endothelial NOS, constitutive in nature and activated by Ca^{2+} and calmodulin; neuronal NOS, also constitutive in nature and activated by Ca^{2+} and calmodulin; inducible NOS, present within macrophages, glial cells, and other cellular elements, Ca^{2+}- and calmodulin-independent but activated by cytokines.[3] In the brain and at the postsynaptic level, NO exhibits a high affinity for the heme group, which it has been suggested, might play a receptor role.[4] The NO binding with the heme of the soluble guanylate cyclase induces the production of cyclic guanosine monophosphate (cGMP). This process, however, is not exclusive since NO can also react with superoxide anions to form peroxinitrites, cysteine radicals to form s-nitrosilated residues, and the Fe–S enzymatic center of the mitochondrial respiratory chain.[5,6] Finally, it is also reported that glutamate, in allowing Ca^{2+} entry, stimulates NO production through N-methyl-D-aspartate (NMDA) receptors, and through a retroinhibitory control upon these receptors NO may regulate its own production.[6]

Regarding the anatomical aspect, several maps describing the brain distribution of the neuronal sets synthesizing NO (NOS antibodies or NADPH-diaphorase labeling) are now available.[7-9] From a general point of view, it first appears that these sets are not widespread in the brain but are occupying well-defined positions and cosynthesize well-known neurotransmitters, such as GABA (cortical interneurons), acetylcholine (hippocampus, hypothalamus, thalamus, nucleus latero dorsal tegmenti, pedunculopontin nucleus), and somatostatine and NPY (olfactory bulb, cerebral cortex, striatum).[7] Concerning monoaminergic neurons, those expressing tyrosine hydroxylase do not contain NOS (substantia nigra, nucleus locus coeruleus, hypothalamus, olfactory bulb), except for a limited neuronal population located in the periaqueductal gray area and the rostroventral tegmentum.[10] An important proportion of the serotoninergic neurons located in the rostral raphe also expresses a NOS activity.[10-12] This last aspect drew our attention, since the involvement of serotonin (5-HT) in sleep triggering and maintenance has been known for several decades, and our research is focused on the study of the sleep–wake cycle mechanisms.[13-15] Thus, if the costorage 5-HT-NO underlies a functional reality, it appears likely that NO might play a part in sleep–wake processes.

Furthermore, since NO is a gaseous and labile compound difficult to measure in tissue biopsies, the great majority of the methods employed to evaluate its concentration *in vivo* are indirect. In this way, citrulline,[16] nitrites, nitrates,[17,18] NOS activity or cGMP[19,20] are now measured as parameters linked to the NO production or efficiency. According to the multiple biological aspects in which this compound is involved, a technology allowing its direct measurement *in vivo* now appears necessary. In this respect, direct measurement of NO *in vivo* can be, indeed, envisaged by using electrochemical methods. Sensors with an active part constituted by a carbon fiber or a platinum wire coated with specific layers are now proposed.[21,22] We prepared an electrochemical sensor[23,24] and here the following are reported:

- Its detailed description
- The test performed either *in vitro* or *in vivo*, supporting the specificity of the measurements
- An applied aspect in which the NO sensor was used, i.e., the brain NO fluctuations occurring throughout the rat sleep–wake cycle.

II. Material and Methods Related to the NO Sensor

A. *NO Sensor Preparation*

A pyrolytic carbon fiber (ϕ = 30 µm, Textron, AVCO, USA) is inserted into a glass pipette (GC100T-10, Clark Electromedical Instruments, Pangbourne, England). The fiber extends beyond the stretched tip by about 500 µm. At this level, the contact "glass-carbon fiber" is sealed with a special glue (resin Sody 33, ESCIL France). To ensure the electrical contact with the carbon fiber, a silver wire (ϕ = 100 µm, A-M systems, Inc., USA), previously dipped in a conductive resin (Elecolit, Radialex, Lyon, France), is inserted into the nonstretched tip of the pipette. Again, at this tip, the wire is sealed to the pipette with an acrylic dental cement (Houmedica International LTD, Surrey, England). At this step, all the sensors prepared are placed in an incubator at 40°C for 12 hours. Just before use, the surface of the carbon fiber is conditioned by applying a triangular current (80 Hz, 2.9 V/20s, 1.3 V/4s, carbon fiber dipped in PBS, phosphate buffered saline). Then, the fiber is successively coated[23,24] (constant voltage = 1.6 V/ 5 × 30 s) with porphyrin–nickel (PN, nickel {nickel–(II)–Tetrakis –(3–OCH3–4–OH-phenyl)–porphyrin}, Interchim) and Nafion® (3.0 V/10 s) (perfluorinated ion-exchange powders, solution aliphatic alcohol, Aldrich). The sensor prepared as stated above can be used immediately or stored at ambient temperature for deferred use for several weeks. (See Figure 3.1.)

B. *Differential Normal Pulse Voltammetric (DNPV) Measurements*

The DNPV method is now also well known and widely used.[25,26] It consists of applying successive double pulses (prepulse and pulse). The prepulse is increasing in amplitude. The current measured is the differential of the current existing between the ends of the pulse and the prepulse. The signal obtained is defined as an oxidation peak and the intensity of the current measured is in the nano-ampere range. Volta-

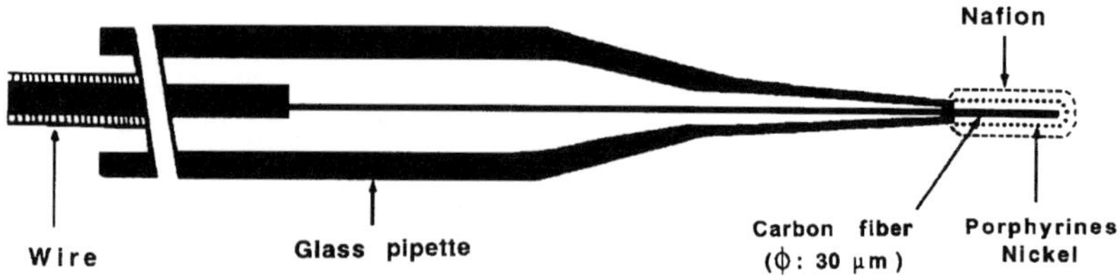

FIGURE 3.1
Sensor used for NO determination: Its active part (carbon fiber, ϕ = 30 μm; length = 500 μm) is coated (electrodeposition) with Porphyrin-Nickel and Nafion. The wire (silver) ensures electrical contact with the carbon fiber.

mmetric measurements (linear potential sweep: 400 to 1350 mV; scan rate: 10 mV/sec; measuring pulse, amplitude: 40 mV, duration: 60 ms) are performed with a pulsed voltammetric unit (Biopulse, Radiometer-Tacussel, France) together with the conventional three-electrode potentiostatic system (working electrode = sensor; reference electrode = Ag/AgCl wire, ϕ = 100 μm, A-M systems, Inc., USA; auxiliary = platinum wire, ϕ = 50 μm, A-M systems, Inc., USA).

C. Preparation of the NO Standard Solutions

NO standard solutions are prepared in anaerobic conditions. For this purpose, 40 ml of deionized water are introduced in a gas-proof chamber and bubbled with argon (HP, Carboxique Française) for 30 min. The oxygen-free water obtained is then bubbled with NO gas (electronic quality: 99.9%, Air Product) for 30 more min. The NO saturated solution (1.98 m*M* at 22°C /*Handbook of Chemistry and Physics*, 1992, page 1607) obtained in this way is then diluted with a deoxygenated PBS solution. Calibration of the electrodes is always performed with fresh solutions under controlled temperature (22°C or 37°C).

III. *In Vitro* Applications

A. Linearity of the Sensor Response

Signals obtained in NO solutions at 22°C and ranging from 5×10^{-7} to 10^{-4} *M* appear at a 650 mV potential and exhibit a linear increase. This is also verified when using the same range of concentrations at 37°C. It must be noticed that for a constant concentration of NO, the signal increases with the temperature. For *in vivo* experiments, it is thus important to calibrate the electrodes in NO solutions at 37°C.

B. Specificity Test

When the active and coated part of the sensor is dipped in PBS solution, the baseline level obtained does not present oxidation peaks. When dipped in NO solutions an

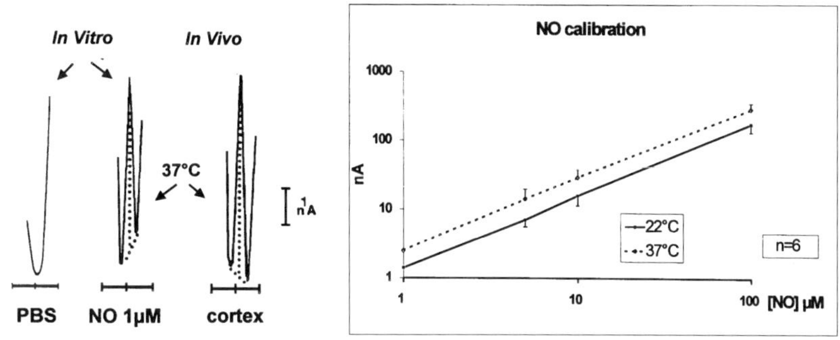

FIGURE 3.2
In vitro PBS: Baseline obtained by use of differential normal pulse voltammetry (DNPV) together with electrodes coated with Porphyrin-Nickel and Nafion and dipped in a phosphate buffered saline solution. *NO 1 μm:* Signal obtained at 650 mV with a coated electrode dipped in a 1 μm NO solution at 37°C; dots indicate the height of the signal and the baseline level. *In vivo* cortex: Signal obtained at 650 mV with the same electrode inserted into the frontal cortex (1st 500 μm) of an anesthetized rat. As compared to the *in vitro* one, it corresponds to an NO extracellular concentration of about 1.3 μ*M*; dots, same remarks as *in vitro*. NO calibration: Variations of the 650 mV (± SEM) signal (Log-Log representation; abscissa: NO concentration; ordinates: peak height) established with 6 electrodes (n = 6) dipped in NO solutions at 22°C or 37°C and increasing in concentration.

oxidation peak is obtained at 650 mV. With an uncoated sensor (without PN and Nafion layers), dipped in a solution of peptides (posthypophyseal extract, Choay), a signal is obtained at 650 to 700 mV.[24] With the coated one, only a baseline level analogous to that yielded by PBS is obtained. This result is due to the repellent properties of the Nafion layer toward the peptide anions present in the solution. The same phenomenon is observed with solutions of nitrites (10 μ*M*), L-citrulline (5 m*M*), and L-arginine (500 μ*M*). Again, it must be noticed that under these conditions the coated electrodes detected a signal peaking at 750 mV only at high concentrations, out of the physiological range. Finally, we also checked *in vitro* that the substances administered *in vivo* do not yield any signal. (See Figure 3.2.)

IV. *In Vivo* Experiments

A. *Pharmacology*

In order to test whether the sensor response *in vivo* varies according to the known effect of the substances injected, an NO donor or an NOS inhibitor are administered intraperitoneally (i.p.) to the anesthetized rat. For this purpose, OFA male rats (250 g, IFFA CREDO-France) are prepared as previously described.[27,28] Briefly, the animals are anesthetized with chloral hydrate (400 mg/kg, i.p., Merck) and after full induction of the anesthesia, they are kept on a stereotaxic frame and their body temperature maintained at 37°C by a homeothermic blanket. The resection of the superficial cutaneous and muscular layers, as well as the bone trepanation, are then performed.

Afterward, the working electrode (NO sensor) is stereotaxically inserted into the frontal cortex (bregma = –2 mm, lateral = 1.5 mm, depth: first 500 μm/Paxinos and Watson atlas[29]) and the auxiliary and reference electrodes placed in contact with the dura. Measurements are taken every 2 min according to the parameters described before and stored on a recorder (Servogor 120, Goerz Instruments, Germany). The signal obtained in the frontal cortex occurs, as *in vitro*, at around 650 mV. After about 30 min it stabilizes and remains unchanged for at least 10 h. According to the size of the active part of the sensor (30 μm), as well as to the calibration curve obtained at 37°C *in vitro*, it might reflect an extracellular concentration of NO of about 1.3 μ*M*.[24] Administration of saline (drug's solvent) does not change its height significantly. The administration of an NO donor (hydroxylamine, 40 mg/kg i.p., Sigma) increases the signal (+100% in about 50 min).[24] On the other hand, the administration of an NOS inhibitor (L-ANA: L-arginine p-nitroanilide, 100 mg/kg i.p., Sigma) is followed by a rapid disappearance (50 min; half-life: 10.5 ± 1.9 min) of the signal.[24] (See Figure 3.3.)

B. NO Fluctuations Throughout the Sleep–Wake Cycle

In compliance with the procedure previously described[24,30] and in order to check how the NO signal fluctuates according to the vigilance states, three rats are equipped, under chloral hydrate anesthesia (400 mg/kg i.p.), with cortical electro-

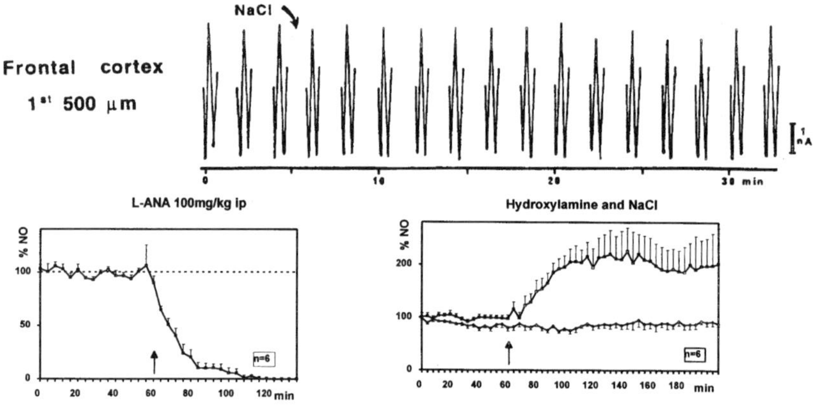

FIGURE 3.3
Frontal cortex, 1st 500 μm: Direct recording obtained *in vivo* with a coated electrode inserted into the cortex of the anesthetized rat. Measurements are performed every two minutes. The signal stability is maintained for up to 10 hours (period assayed). NaCl does not change the baseline level. Abscissa: time in minutes (min); nA: nano-Ampere. L-ANA (L-Arginine-p-Nitro-Anilide, NOS inhibitor, 100 mg/kg i.p.): Mean evolution (± SEM) of the signal height (ordinates: % of the peak height) induced by L-ANA administration (arrow, n = 6 rats); its complete disappearance is obtained in about 50 minutes. Statistics: abscissa from 68 to 80 min $p < 0.01$; above 80 min, $p < 0.001$ (ANOVA followed by a multiple range test). Hydroxylamine: mean evolution (± SEM) of the effect obtained (ordinates: +100% of the peak height; n = 6 rats) after administration of the NO donor (arrow, i.p.). Statistics: abscissa from 80 to 88 min, $p < 0.05$; above 88 min $p < 0.001$ (ANOVA followed by a multiple range test).

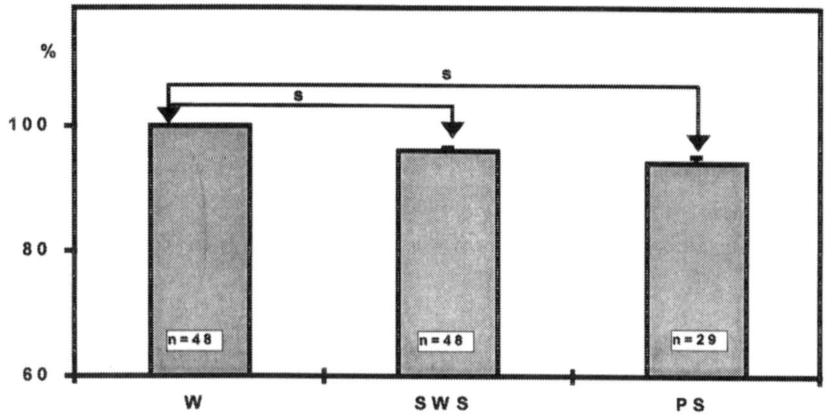

FIGURE 3.4
Mean variations of the NO signal occurring in the frontal cortex of the unanesthetized rat according to the vigilance state: the highest values are measured during the waking state (W). From this state (referenced 100%), the signal height decreases during slow-wave sleep (SWS: 6%; SWS/W: n = 48 cycles, Student's t-test, $p < 0.004$) and even more during paradoxical sleep (PS: 9%; PS/W: n = 29 cycles, $p < 0.008$).

encephalographic (EEG) and neck muscle (EMG) electrodes. Reference and auxiliary electrodes necessary for voltammetric measurements are also inserted in contact with the dura. Finally, a micromanipulator,[30] allowing the NO sensor insertion in the unanesthetized animal (recording sessions), is stereotaxically implanted into the frontal cortex (bregma = –2 mm, lateral = 1.5 mm, depth = first 500 μm). All the electrodes are soldered to a pair of subminiature five-pin connectors and the entire assembly is cemented to the skull using a dental acrylic cement. The animals are then placed in individual home-cages at 24 ± 1°C and maintained under a 12 h/12 h light–dark cycle with food and water *ad libitum.* Combined voltammetric and sleep polygraphic recordings begin 10 days after surgery. Polygraphic recordings are automatically interrupted only for DNPV measurements. Variations of the NO peak height occurring during either slow-wave sleep (SWS) or paradoxical sleep (PS) are expressed in percentage as compared to the waking state (W, 100%). In the frontal cortex, the highest values of the signal are measured during waking (100%). During SWS and PS, its height decreases progressively (–6% during SWS / W; –9% during PS / W).[24] (See Figure 3.4.)

V. Interpretation and Limits of the Method and Results

The electrochemical sensor described is well adapted for approaches combining electrochemical measurements and behavioral studies requiring freely moving animals. Its specificity toward NO can be argued according to the following relevant facts:

- A voltammetric signal peaking at around 650 mV is obtained either in standard solutions of NO or in brain tissue
- The compounds interfering with NO production *in vivo*, i.e., nitrites, nitrates, L-arginine and L-citrulline, do not yield a voltammetric signal at physiological concentrations; only at extraphysiological concentrations is a signal obtained at 750 mV with nitrites and L-arginine; this signal is above 650 mV
- Peptides, present *in vivo,* are not detected at physiological concentration by the coated sensor since Nafion membrane exerts an efficient repellent role toward the peptide anions
- Whether PN membrane is necessary is now questioned by experiments performed *in vitro*;[23] according to our own experience, both layers confer stability and sensitivity to the sensor used in long-term chronic conditions
- Catecholamines or 5-hydroxyindolamines, also present *in vivo*, do not contribute to the 650 mV peak since they oxidize, respectively, at 100 and 300 mV[27,28]
- The 650 mV signal increases after administration of hydroxylamine (an NO donor) and disappears completely after administration of L-ANA (an NOS inhibitor).

While the above experimental facts ensure that the signal measured by our sensor in the rat brain might be dependent upon the NO fraction present in the extracellular space, attention must be addressed to the possibility of pitfalls, such as:

- The use of substances electroactive at 650 mV
- Temperature variations induced by substances inducing secondary passive changes in the signal height
- Alteration in the Nafion membrane impermeability allowing access to the active surface of the sensor by compounds like peptides; this aspect can be checked at the end of each experiment by administering an NOS inhibitor which must completely suppress the 650 mV signal.

Finally, concerning the data obtained in the frontal cortex of the sleeping or waking rat, it must be noticed that mild but significant variations are measured throughout the sleep–wake cycle by our sensor. Its sensitivity thus appears to be well adapted for the detection of the concentration changes occurring in physiological situations where important changes cannot be expected. Our data also report that the fraction of NO present in the intracellular space of the frontal cortex is higher during W than during SWS and PS. This fact allows the suggestion that NO might be more actively produced during W. However, here it might be emphasized that within the cortex several NO sources exist, i.e., local GABAergic interneurons, axonal nerve endings coming from the basal hypothalamus (cholinergic) or the rostral raphe (serotoninergic).[7] The NO release observed could thus represent the average variations of these three sources at least, which might release NO separately with different relationships toward the sleep–waking cycle. According to the present knowledge, it is clear that further experiments are still necessary to clarify the role or the roles exerted by NO in sleep processes.

Acknowledgments

This work was supported by INSERM-U52, CNRS-ERS5645, and SYNTHELABO (Donation Veille-Sommeil). We also thank C. Limoges for improving the English text and G. Debilly for computerizing the figures.

References

1. Snyder, S. and Bredt, D., Les fonctions biologiques du monoxyde dazote, *Pour la Science,* 177, 70, 1992.
2. Bredt, D. and Snyder, S., Nitric Oxide: a physiologic messenger molecule, *Annu. Rev. Biochem.,* 63, 175, 1994.
3. Mayer, B., Biochemistry and molecular pharmacology of nitric oxide synthases, in *Nitric Oxide in the Nervous System,* Vincent, S., Ed., Academic Press, London, 1995, 21.
4. Koesling, D., Humbert, P., and Schultz, G., The NO receptor: characterization and regulation of soluble guanylyl cyclase, in *Nitric Oxide in the Nervous System,* Vincent, S., Ed., Academic Press, London, 1995, 43.
5. Manzoni, O., Prezeau, L., Marin, P., Deshager, S., Bockaert, J., and Fagni, L., Nitric oxide-induced blockade of NMDA receptors, *Neuron,* 8, 653, 1992.
6. Shuman, E. M., 1995. Nitric oxide signalling, long-term potentiation and long-term depression, in *Nitric Oxide in the Nervous System,* Vincent S., Ed., Academic Press, London, 1995, 125.
7. Vincent, S. R. and Kimura, H., Histochemical mapping of nitric oxide synthase in the rat brain, *Neuroscience,* 46, 755, 1992.
8. Vincent, S. R., Localization of nitric oxide neurons in the central nervous system, in *Nitric Oxide in the Nervous System,* Vincent, S., Ed., Academic Press, London, 1995, 83.
9. Rodrigo, J., Springall, D. R., Uttenthal, C., Bentura, M. L., Abadia-Molina, F., Riveros-Moreno, V., Martinez-Murillo, R., Polak, J. M., and Moncada, S., Localization of nitric oxide synthase in the adult rat brain, *Phil. Trans. R. Soc. Lond.,* 345, 175, 1994.
10. Johnson, M. D., Localization of NADPH diaphorase activity in monoaminergic neurons of the rat brain, *J. Comp. Neurol.,* 332, 391, 1993.
11. Wotherspoon, G., Rattray, A. M., and Priestley, J. V., Serotonin and NADPH-diaphorase in the dorsal raphe nucleus of the adult rat, *Neurosci. Lett.,* 173, 31, 1994.
12. Wang, Q. P., Guan, J. L., and Nakai, Y., Distribution and synaptic relations of NOS neurons in the dorsal raphe nucleus: a comparison to 5-HT neurons, *Brain Res. Bull.,* 37, 177, 1995.
13. Jouvet, M., Biogenic amines and states of sleep, *Science,* 163, 32, 1969.
14. Cespuglio, R., Houdouin, F., Oulerich, M., El Mansari, M., and Jouvet, M., Axonal and somato-dendritic modalities of serotonin release: their involvement in sleep preparation, triggering and maintenance, *J. Sleep Res.,* 1, 150, 1992.
15. El Kafi, B., Leger, L., Seguin, S., Jouvet, M., and Cespuglio, R., Sleep permissive components within the dorsal raphe nucleus in the rat, *Brain Res.,* 686, 150, 1995.
16. Calberg, M., Assay of neuronal nitric oxide synthase by HPLC determination of citrulline, *J. Neurosci. Meth.,* 52, 165, 1994.

17. Ohta, K., Araki, N., Shibata, M., Hamada, J., Komatsumoto, S., Shimazu, K., and Fukuuchi, Y., A novel in vivo assay system for consecutive measurement of brain nitric oxide production combined with microdialysis technique, *Neurosci. Lett.*, 176, 165, 1994.
18. Luo, D., Knezevich, S., and Vincent, S. R., N-methyl-D-aspartate-induced nitric oxide release: an in vivo microdialysis study, *Neuroscience,* 57, 897, 1993.
19. Vallebuona, F. and Raiteri, M., Monitoring of cyclic GMP during cerebellar microdialysis in freely-moving rats as an index of nitric oxide synthase activity, *Neuroscience,* 57, 577, 1993.
20. Luo, D., Leung, E., and Vincent, S. R., Nitric oxide-dependent efflux of cGMP in rat cerebellar cortex: an in vivo microdialysis study, *J. Neuroscience,* 14, 263, 1994.
21. Shibuki, K., An electrochemical microprobe for detecting nitric oxide release in brain tissue, *Neurosci. Res.,* 9, 69, 1990.
22. Malinski, T., Mesaros, S., and Tomboulian P., Nitric oxide measurement using electrochemical methods, in *Methods in Enzymology,* Packer, Lester, Ed., Academic Press, New York, 1996, 268, 58.
23. Fabre, B., Bidan, G., Cespuglio, R., and Burlet, S., French Patent CEA/INSERM, Brevatome, n° 94 10290, Electrode et capteur de détection in vivo du monoxyde d'azote et de ses dérivés et procédé de détection ampérométrique in vivo du monoxyde d'azote et de ses détivés, 1994.
24. Cespuglio, R., Burlet, S., Marinesco, S., Robert, F., and Jouvet, M., NO voltammetric detection in the rat brain: variations of the signal throughout the sleep-waking cycle, *C. R. Acad. Sci. Paris / Life Sciences,* 319, 191, 1996.
25. Rivot, J. P., Cespuglio, R., Puig, S., Jouvet, M., and Besson, J. M., In vivo electrochemical monitoring of serotonin in spinal dorsal horn with nafion-coated multi-carbon fiber electrodes, *J. Neurochemistry,* 65, 1257, 1995.
26. Suaud-Chagny, M. F., Cespuglio, R., Rivot, J. P., Buda, M., and Gonon, F., High sensitivity measurement of brain catechols and indoles in vivo using electrochemically treated carbon-fiber electrodes, *J. Neurosci. Meth.,* 48, 241, 1993.
27. Cespuglio, R., Sarda, N., Gharib, A., Houdouin, F., and Jouvet, M., Voltammetric detection of the release of 5-hydroxyindole compounds throughout the sleep-waking cycle of the rat, *Exp. Brain Res.,* 80, 121, 1990.
28. Houdouin, F., Cespuglio, R., and Jouvet, M., Effects induced by the electrical stimulation of the nucleus raphe dorsalis upon hypothalamic release of 5-hydroxyindole compounds and sleeEfaut111Efaut111Efaut111Efaut111p parameters in the rat, *Brain Res.,* 565, 48, 1991.
29. Paxinos, G. and Watson, C.,*The Rat Brain in Stereotaxic Coordinates,* Academic Press, London, 1986, 245.
30. Louilot, A., Serrano, A., and D'Angio, M., A novel carbon fiber implantation assembly for cerebral voltammetric measurements in freely moving rats, *Brain Res.,* 41, 227, 1987.

Chapter 4

Sleep Regulatory Substances: Change in mRNA Expression Linked to Sleep

Zutang Chen and James M. Krueger

Contents

0-8493-3361-X/98/$0.00+$.50

I. Introduction

The accumulation of sleep regulatory substances (SRS) in cerebrospinal fluid (CSF) during prolonged wakefulness provides very strong support for the hypothesis that sleep is regulated, in part, by humoral agents.[reviewed 1-4] Many substances can affect sleep, although only a handful of humoral agents are strongly implicated in sleep regulation. The list includes tumor necrosis factor-α (TNFα), interleukin-1β (IL-1β), growth hormone releasing hormone (GHRH), prostaglandin D_2 and adenosine for nonrapid eye movement sleep (NREMS), and vasoactive intestinal peptide and prolactin for REMS.[reviewed 1,4,5] All putative SRSs thus far identified have other biological activities not directly tied to sleep. For example, IL-1 and some prostaglandins are pyrogenic, yet body temperature decreases upon entry into NREMS.[6-8] Similar problems of specificity confront the humoral and neuronal regulation of all physiological functions. Any manipulation that alters sleep can also alter other physiological parameters, e.g., sleep deprivation is associated with changes in body temperature, food intake, and endocrine function. It is therefore our opinion that independent approaches affecting sleep should be used to determine sleep-specific changes in either SRSs or in neuronal networks. Thus, if one is interested in determining sleep-specific localized changes in an SRS mRNA, only changes mutually induced by independent sleep-altering methods can be considered candidates for specific SRS sleep mechanisms.

There are several important reasons why one should determine sleep-driven changes in SRSs. For some SRSs, e.g., IL-1, TNF, GHRH, there is now considerable direct evidence that they form part of the causal chain of events leading to sleep; determination of where and by how much they change in brain allows one to investigate causal sleep mechanisms directly. In contrast, recordings from single cells during sleep–wake cycles will always remain correlational since one cannot know if the cell recorded from is causally related to sleep. Second, localization of sleep-driven changes in SRSs will help decipher how the brain is organized to produce sleep. Third, as our knowledge of SRSs expands there will likely be practical applications useful in treating sleep disorders. Fourth, it is likely that knowledge of SRSs and their mechanisms of sleep induction will lead to testable theories of sleep function.

II. The Role of IL-1β

By way of example this essay will focus on IL-1β. The experimental approaches and problems associated with determination of sleep-linked changes in IL-1β also apply to other SRSs. Before undertaking the difficult task of determination of sleep-linked changes in an SRS there should be extensive data linking the SRS to sleep. Very briefly, the evidence linking IL-1 to sleep is as follows: exogenous IL-1β induces increases in NREMS in rabbits, rats, cats, mice, and monkeys and induces sleepiness in humans.[reviewed 1,9-15] Inhibition of IL-1β using antibodies to IL-1β, or a peptide fragment of the IL-1 soluble receptor, or the IL-1 receptor antagonist

(IL-1RA) inhibits spontaneous sleep.[16-19] Further, substances that inhibit the actions or production of IL-1β also inhibit sleep, e.g., interleukin-10 (IL-10) and prostaglandin E_2.[20,21] IL-1 plasma levels and IL-1 bioactivity in cerebrospinal fluid vary in phase with the sleep–wake cycle, with highest levels occurring at the onset of sleep.[22-26] The IL-1 family of molecules including IL-1β, IL-1 receptors, the IL-1RA and the IL-1 receptor accessory protein (IL-1AP) are all constitutively expressed in normal brain.[reviewed 1] IL-1β mRNA levels in rat brain exhibit a diurnal rhythm in the hypothalamus, hippocampus, and cortex with highest levels present just after lights are turned on, the period of maximum sleep in rats.[27] Further, IL-1β mRNA increases in the hypothalamus and brain stem during sleep deprivation.[28] IL-1β affects a variety of neurotransmitter systems involved in sleep regulation, e.g., serotonin, acetylcholine, histamine, and GABA.[reviewed 1,29] For example, IL-1β, via an IL-1 receptor and the $GABA_A$ receptor potentiate Cl^- permeability; this mechanism is thought to be involved in electroencephalographic (EEG) synchronization.[30,31]

IL-1β is a polypeptide with a molecular weight of 17 kD that has autocrine, exocrine, and endocrine roles. The IL-1 family has nine known members. There are three IL-1 ligands: IL-1α, IL-1β, and the IL-1RA; these members share limited amino acid homology although they share a higher homology in their predicted three-dimensional topologies; all bind to IL-1 receptors.[32] The IL-1RA competes with IL-1α and IL-1β for binding sites and, as its name implies, it antagonizes the actions of IL-1α and IL-1β.[e.g., 33] Two IL-1 receptors have been identified, Type I and Type II. They possess three extracellular immunoglobulin-like domains, limited homology, and different binding characteristics.[reviewed 23,34,35] The two receptors have different functions; the Type I is the signal transducing receptor, while the Type II is thought to be a decoy receptor having a truncated nonsignaling intracellular domain.[36] Recently an IL-1 receptor accessory protein (IL-1AP) has been identified; it has limited homology with Type I and Type II IL-1 receptors and is found in brain.[37,38] The IL-1AP forms a complex with the Type I receptor and either IL-1α or IL-1β, but not with the IL-1RA. The IL-1AP increases the binding affinity of IL-1β for the Type I receptor, and it is important for signal transduction.[38,39] An IL-1 receptor-associated protein kinase (IRAK), which activates nuclear factor kappa B (NF-kB), has been cloned.[40] Another IL-1 receptor-related protein closely related to the IL-1 receptors has recently been described; it is expressed in brain and is associated with the cerebral vasculature.[41] IL-1α and IL-1β are first made as precursor molecules; mature forms are released from cells. The IL-1α precursor is biologically active, while pre-IL-1β requires processing by the IL-1β-converting enzyme (ICE) which cleaves pre-IL-1β at two sites, thereby producing the mature 17 kD IL-1β fragment. Finally, an alternatively processed cDNA for the rat IL-1 Type II receptor which only encodes the extracellular domain of the receptor was described.[42] This soluble IL-1 receptor probably acts to bind and thereby reduce the effective concentration of IL-1.

This is a complex family of molecules; one may ask: why not just measure levels of IL-1β and components of the functional gene group involved in NREMS regulation using RIA and ELISA kits? Although such determinations should remain a long-term goal, there are several reasons why such experiments are premature. Use of cytokine kits for determination of levels of cytokines has a history of complicating factors. For example, we have shown that an IL-1 receptor fragment displaces the standard curve in an IL-1β kit.[18] Thus, if sleep-linked changes in the IL-1 family of molecules include changes in soluble receptor levels, measured IL-1β levels could be affected. Many, if not all, of the RIA and ELISA kits available are currently insufficiently sensitive to measure brain levels of IL-1β. Further, if the antibodies used in ELISA assays are used in a Western blot of brain extracts, multiple bands are observed. This leads one to question the specificity of the antibodies. We do not yet know which members of the IL-1 family are affected by sleep–wake cycles or where such sleep–wake cycle-induced changes occur in brain. Further, RIA and ELISA kits are available for only humans or mice for only three of the nine members of the IL-1 family; in theory the IL-1 system could be regulated by changes in any one of the IL-1 family members. Finally, using nucleic acid chemistry provides greater specificity.

The strategies used to localize sleep-linked changes depend upon what one assumes about how sleep is regulated. While there are several neuronal networks involved in NREMS regulation, including the hypothalamic preoptic-basal forebrain area and thalamo-cortico circuits, there are no demonstrated *necessary* neuronal networks for NREMS. It is thus prudent to begin experiments with the *a priori* assumption that sleep-linked changes in an SRS may occur anywhere in the brain and may not be localized to traditional neuronal sleep–wake sites focused on by others. A second important consideration that influences localization strategy is demonstrating that any change observed is in fact tied to sleep. One cannot conclude that changes observed after a single manipulation of sleep, e.g., sleep deprivation, are sleep-linked since they could be associated with other variables that change with sleep, e.g., brain temperature.[7,8] A premature focus on any specific neuronal network therefore could be devastating in the long-term. Third, it is possible that sleep-linked changes of different members of the SRS family are differentially localized. Finally, a theoretical consideration tempers one's assumptions about what is known about localization of sleep regulation. The minimal component of brain that is capable of sleep is unknown. We have proposed that sleep is a fundamental property of neuronal groups and that sleep, and changes in the humoral agents that drive sleep, begin at that level.[43,44] Further, Pigarev and colleagues have shown that sleep develops asynchronously in cortical areas.[45] The distinction of whether sleep-linked neuronal networks coordinate or initiate sleep has not been made. Since there is no known brain lesion that results in complete and permanent loss of NREMS (hundreds of such studies are in the literature), current evidence favors the notion that known sleep-linked networks coordinate sleep. This issue is important since it has direct bearing on sleep mechanisms and sleep function. Such considerations indicate that it is best to begin localization studies by examining relatively large areas.

III. Specific Methods for mRNA Quantification

A. General Approaches

An important aspect in the determination of gene expression related to sleep regulation is to monitor the changes of mRNA levels of the specific SRS gene of interest.

The techniques for quantification of mRNA levels include Northern blotting, dot and slot blotting, solution hybridization, RNase protection, nuclear run-on transcription, and reverse-transcriptase polymerase chain reaction (RT–PCR). Dot and slot hybridization methods are used for quick and coarse RNA measurements; RNA samples are directly spotted through a filtration manifold blotted onto a dry solid support (usually nitrocellulose or nylon membrane) and then hybridized with DNA or RNA probes. These methods are similar to Northern blot hybridization except RNAs are not separated by electrophoresis. Solution hybridization actually is a part of the RNase protection assay. After radiolabeled probe and targeted RNA are hybridized, radioactivity of the hybrid is determined. We will focus on Northern blotting, RNase protection, and RT–PCR in this chapter.

Molecular study of somnogenic-related molecules requires use of their cDNA clones. These clones can be provided by the lab where the cDNAs were originally cloned or cloning by PCR the probe of interest oneself. The PCR product of the cDNA of interest can be directly subcloned into a A–T clone vector such as pGEM-T (Promega, Madison, WI) if Taq polymerase is used in PCR, because an extra A is usually added into the sequence at the 3′-end. In contrast, if Pfu polymerase is used in PCR, a blunt-ended PCR product is generated, so that it needs to be subcloned into a blunt-ended linearized vector such as pCR-Script (Stratagene, La Jolla, CA). The direction of the cloned PCR product in these constructs needs to be determined by restriction mapping. Alternatively, two different restriction sites can be designed into the ends of the 5′ and 3′ primers, respectively, so that the PCR product can be subcloned into the corresponding sites of the vector in an oriented manner. For this directional-oriented subcloning, the restriction sites should be unique and not exist in the internal sequence of the cDNA clone of interest. Several suitable *E. coli*-derived cloning vectors are provided from different venders for *in vitro* transcription from a cloned recombinant DNA insert. The polylinker in these vectors is flanked by T3, T7, or SP6 RNA polymerase promoter sequences. First a cDNA sequence representing the mRNA of interest needs to be subcloned into one of those vectors. When it is desirable to transcribe the insert sequence, the plasmid is linearized with a restriction enzyme at the end of, or within, the inserted sequence. When the mixture of labeled and cold NTPs (N stands for A, T, C, and G) are provided, the RNA polymerase recognizes its own promoter sequence and starts to transcribe the downstream inserted sequence. Discrete sense run-off transcripts will be obtained by transcription from the 5′ upstream promoter, whereas radiolabeled antisense RNA probes will be transcribed from the 3′ downstream promoter and can be used for Northern and Southern blots, *in situ* hybridization, and RNase protection assay. The typical yield is 5 to 10 μg RNA per μg of plasmid DNA template.

B. RNA and RNA Extraction

RNA is found in the nucleus, cytoplasm, and mitochondria of eukaryotic cells. Total cytoplasmic RNA consists mainly of ribosomal RNA (rRNA), transfer RNA (tRNA), and messenger RNA (mRNA). The majority of total RNA is 28S and 18S rRNA, whereas mRNA is only 1 to 2% of total RNA. Approximately 500,000 mRNA molecules have been estimated to be in one cell. The key point in obtaining a nondegraded RNA preparation is to effectively minimize RNase activity. Although there are many RNA extraction protocols from different sources, during cell lysis, quick and effective inactivation and/or digestion of endogenous RNases is essential. Since RNases are stable, heat-resistant (normal autoclaving does not denature these enzymes), and widespread, one should use RNase-free plasticware for RNA preparation and storage. Diethyl pyrocarbonate (DEPC) and other RNase inhibitors are recommended for use before or during the preparation.

Two major RNase denaturants, phenol and guanidinium thiocyanate, are commonly used in RNA isolation protocols.[46] Some commercial kits provide a single-step purification procedure using both denaturants. RNAzol B and STAT-60 (Tel-Test, Friendswood, TX) work satisfactorily for total RNA extraction. Depending upon the tissue, 1 gram of tissue may yield 2 to 8 mg total RNA or 20 to 100 μg mRNA, and 10^8 cells can yield 1 to 2 mg total RNA or 10 to 40 μg mRNA. The minimum amount of tissue to be used for RNA extraction can be as low as 2.5 mg or 10^5 cells.

Many applications for determining RNA levels are performed using total RNA. The advantages of using total RNA are that the extraction procedure is straightforward, and total RNA is a little more resistant to RNases because mRNAs are protected by the excess of rRNA. However, since isolated poly(A)+ RNA is selectively enriched in mRNA, greater sensitivity of detection can be obtained by using mRNA in the application. Detection level from mRNA is 0.0002% compared to 0.02% from total RNA. Heteronuclear RNA, the precursor of mRNA, is transcribed in the nucleus. After processing, intron sequences have been spliced, and the mRNA has been polyadenylated [a poly (A)+ tail is added to the 3′ end of an mRNA]. The principle in isolating mRNA is to use the poly (A)+ tail found in all mRNA. Several poly dT constructs conjugated to some type of solid material, including magnetic beads, agarose, etc., have been developed by different companies for extracting mRNA. After the binding of mRNA to poly dT conjugated to a solid support, followed by washing away unbound RNAs, the mRNA can be eluted using heat or high salt. In general, 10^6 cells or 100 mg tissue can yield 1 mg total RNA or 10 μg mRNA.

C. Specific Methods: Cloning of Rat IL1-β Coding Sequence and Construction of pBIL1-βCS (IL1-β Coding Sequence) and pBIL1-βCStrunc (Truncated IL1-β Coding Sequence), and Their Usefulness

We use the following strategies for cloning the IL1-β coding region by the PCR method; similar procedures can be used for any known cDNA clone. The rat IL1-β cDNA sequence is obtained via the internet by "genebank text searching" under the address http://www.ncbi.nih.gov/web/search/index.html. The sequence shows that IL1-β protein coding region is from first methionine codon 5′–ATG — at position 77 to the stop codon — TAA-3′ at position 883. The restriction map of the IL1-β sequence was determined by DNAsis (Hitachi Software Engineering Co., Ltd.). pBluescript (Stratagene, La Jolla, CA) was chosen as a cloning and expression vector, because the multicloning site is flanked by T3 and T7 RNA polymerase promoters allowing transcription of both sense and antisense RNA. SacI and XhoI were used as the cloning sites because of their absence from the IL1-β cDNA sequence.

1. Primer Design and PCR Cloning

The upstream PCR cloning primer composed of 15 bases from the first methionine codon **ATG** at the position of 77 to 91, tagged with 3 bases plus a SacI restriction sequence at the 5′ end of the primer, reads as: 5′–ACTGAGCTC**ATG**GCAACT-GTCCCT–3′. The downstream antisense primer includes the 15 bases from position 869 to the stop codon **TAA** at position 883. This primer is tagged by an XhoI site with a additional three bases at the 3′ end of the primer; it reads 5′–CCCGTGTCT-TCC**TAA**CTCGAGCTA–3′. However, the downstream antisense primer is complementary, 3′–GGGCACAGAAGG**ATT**GAGCTCGAT–5′, and starts at the 5′ end, so the real primer will be read as 5′–TAGCTCGAG**TTA**GGAAGACACGGG–3′. Underlined are the restriction sites and bold letters stand for the start or stop codon sequences. The addition of three bases to each restriction site ensures the restriction enzyme digestion.

After a coding sequence of IL1-β with the restriction sites is synthesized by PCR from rat cDNA, both vector and PCR product are digested with SacI and XhoI. The insert and vector are ligated by T4 DNA ligase at 4°C overnight yielding the plasmid pBIL1-βCS, standing for the IL1-β coding sequence. The plasmid is amplified in *E. coli* strain DH5α (Gibco BRL, Gaithersburg, MD). The correct sequence is confirmed by DNA sequencing.

2. Probe Preparation and In Vitro Transcription

Probes used to hybridize a targeted mRNA can be synthetic RNA or DNA labeled with radioactive (^{32}P, ^{33}P, ^{35}S, etc.) or nonradioactive (e.g., biotin, Digoxigenin, Boehringer, Mannheim) materials depending upon utilization purpose and personal preference.

a. Labeled DNA probes

There are several strategies to incorporate radioactive labeled phosphate into a DNA substrate.[46] 5′–end labeling is carried out by transferring (γ–^{32}P)phosphate from ATP to the 5′ dephosphorylated DNA sequence by nucleotide kinase. 3′–end labeling is done by adding a mononucleotide from (γ–^{32}P) dNTP, to the 3′-hydroxyl terminus of ssDNA or dsDNA, accompanied by the release of inorganic pyrophosphate. End labelings of oligonucleotides adds only one ^{32}P to each molecule, so they will only give a weak signal relative to the other more efficient method of incorporation of labeled dNTPs into a duplex DNA by nick translation or random prime labeling. DNA templates for labeling can be a plasmid containing an insert of the DNA sequence of interest, or more precisely, a cDNA insert from the plasmid or a PCR product. The DNA probe is then synthesized from the template by DNA polymerase (Klenow) when radiolabeled (or other labels) dNTP and primers are provided. Random DNA 6mer or 9mer are the universal primers used for any DNA templates larger than 400 bp.

b. Labeled RNA probes and RNA in vitro transcription

Transcribed from the 3′ end of the pBIL1-βCS by T7 RNA polymerase, using SacI linearized pBIL1-βCS plasmid as template, a full-length radiolabeled antisense probe (807 nt) is produced. When pBIL1-βCS is linearized by digestion with BamH1 at position 680, a short probe with 203 nt is synthesized. In contrast, if the plasmid is cut with XhoI, then transcribed with T3 RNA polymerase from the 5′ end, a full coding sequence with 807 nt RNA will be produced. A 604 nt RNA is produced from the BamHI cut plasmid transcribed with T3 RNA polymerase. Known amounts of sense RNA can be used as an RNA quantitative standard, whereas labeled antisense RNA can serve as probes for Northern blot and RNase protection assays.

For the RT–PCR internal control, a synthesized mutant IL1-β RNA with a deletion within the span of a pair of PCR primers is obtained from pBIL1-βCStrunc, a truncated pBIL1-βCS with a deletion of a 217 bp PstI fragment. pBIL1-βCStrunc is constructed by digestion of pBIL1-βCS with the PstI restriction enzyme. A 217 bp PstI fragment (from 450 to 667 nt) and the rest of plasmid containing the entire vector and truncated IL1-β sequences are separated on an agarose gel by electrophoresis. The larger fragment of the DNA is recovered from the sliced agarose by Geneclean II Kit (Bio 101, Vista, CA), religated with DNA ligase, and amplified in *E. coli*. The mutant RNA is then transcribed from the XhoI linearized plasmid with T3 RNA polymerase from the 5′ end T3 promoter. A known amount of this mutant RNA is added to wild-type sample RNA, and these are amplified together by RT–PCR. The wild-type RNA and shorter RNA RT–PCR products are quantitatively compared.

3. Northern Blot Analysis

This approach for RNA detection is fairly standard, and there are numerous detailed descriptions from various sources.[46] According to several references, RNA detection levels using Northern blot are in the range of pg levels or 10^5 to 10^7 copies of the molecule.[47] Ten to thirty μg of total RNA for abundant messages (~0.1% of mRNA) or 0.5 to 3 μg poly(A)+ RNA for rare signals are reasonable amounts to be separated on 1.2% agarose–6% formaldehyde denaturing gels. The ribosomal bands (if total RNA is used) can be visualized by staining the gel with ethidium bromide at a concentration of 0.5 μg/ml H_2O. The RNA is then transferred to reinforced nitrocellulose membranes with 20X SSC (3 *M* NaCl, 0.3 *M* sodium citrate, pH 7.0) by capillary elution, and the RNA on the membrane is crosslinked by a UV-crosslinker (Stratagene, La Jolla, CA). The attachment of denatured RNA to the nitrocellulose is presumed to be irreversible so that the blot can be hybridized sequentially with a series of probes. Prehybridization of the blot in the presence of denatured herring sperm DNA at 42°C for 2 hours is followed by hybridization with 0.1 μg probe at 2×10^8 dpm/μg in the same buffer for 16 to 24 hours. QuickHyb solution (Stratagene,

La Jolla, CA) with 10 min prehybridization and 1 hour hybridization can also be used. After stringency washing with 0.1X SSC and 0.1% SDS at 42°C, the RNA can be visualized by exposure of the hybridized blot to X-ray film. The sample RNA quantification is carried out by comparison to the intensity of a known amount of synthesized RNA fragment, usually different in size from the wild-type RNA. According to our experiences and calculations, 10^6 to 10^7 copies of mRNA or 10 to 100 pg mRNA can be detected using Northern blot analyses. If one works with a limited amount of tissue with low abundant messages, RNase protection assay or RT–PCR can be used to dramatically increase the signal detection threshold.

4. *RNase Protection Assay*

Since hybridization of free sample RNA to a probe in a liquid environment is more efficient than hybridization to bound RNA on a solid support, the RNase protection assay increases the detection level by 1 to 2 orders of magnitude. Sample RNA (0.5 to 10 μg), 20 pg short form (604 nt) synthetic IL-1β RNA, and 300 to 600 pg (2×10^6 dpm) (α–^{32}P)-labeled antisense RNA probe are mixed in 30 μl hybridization buffer (Boehringer Mannheim, Indianapolis, IN). Equilibrium of hybridization is reached by 4 hrs at 42°C. The free RNA is digested with 3.5 μg RNaseA and 25 units of RNaseT1 in 350 μl digestion buffer at 37°C for 15 min. The protected RNA:RNA hybrids are coprecipitated with tRNA and then separated on 4% denatured (7 *M* urea) polyacrylamide gel with 1X Tris-borate/EDTA electrophoresis buffer (TBE buffer).[46] The protected fragment of sample RNA should be equal in length to the probe, 807 nt; and that from the internal standard should be 604 nt (the RNA is transcribed from 77 to 680, the BamHI site). The nucleotide bands can be visualized under the UV light after the gel is stained with ethidium bromide (0.5 μg/ml). Or autoradiography is carried out by exposing the dried gel on the 3 MM filter to X-ray film.

5. *RT–PCR with Internal and External Controls*[48]

RT–PCR is the most sensitive method for detecting low abundant mRNAs. Since RNA present is amplified many times, a little error is also amplified many times, which may give a false result without the proper controls. One microgram of total RNA from the brain tissue and a known amount (pg) of synthetic mutant IL1-β RNA (for an exogenous internal standard) are mixed prior to the RT–PCR. First strand cDNAs from both authentic IL1-β RNA and mutant IL1-β RNA are reverse transcribed with 1 m*M* of the same downstream probe by 200 U SuperScript Reverse Transcriptase in a final volume of 20 μl according to the manufacturer's instruction (Gibco BRL, Gaithersburg, MD). The PCR is performed in a final volume of 50 μl containing 2 μl of 3.5X diluted cDNA from the above preparation. In order to increase the sensitivity of the detection, 1 μl of 100X diluted (α–^{32}P) dCTP is added to each reaction. The RT–PCR product from the authentic IL-1β RNA primed at position 143 and 711 will be 569 nt, which is 217 bp longer than that from the internal standard RNA due to the deletion of the PstI fragment from 450 to 667 nt.

Alternatively, a known amount of β-actin RNA can be used as endogenous internal control with its own primers parallel with IL-1β RNA's RT–PCR in the same tube. Ten μl of PCR product is analyzed on a horizontal agarose gel. After electrophoresis, the gel is exposed to a photographic plate and the bands are cut from the gel and dpm are determined by liquid scintillation. The amount of RNA from the sample is estimated by comparison with a known amount of external or internal standards.

References

1. Krueger, J. M. and Majde, J. A., Microbial products and cytokines in sleep and fever regulation, *Crit. Rev. Immunol.*, 14, 355, 1994.
2. Borbély, A. A. and Tobler, I., Endogenous sleep-promoting substances and sleep regulation, *Physiol. Rev.*, 69, 605, 1989.
3. Inoué, S., *Biology of Sleep Substances*, CRC Press, Boca Raton, Florida, 1989.
4. Krueger, J. M., Obál, Jr., F., Opp, M., Toth, L., Johannsen, L., and Cady, A. B., Somnogenic cytokines and models concerning their effects on sleep, *Yale J. Biol. Med.*, 63, 157, 1990.
5. Roky, R., Obál, Jr., F., Valatx, J.-L., Bredow, S., Fang, J., Pagano, L. P., and Krueger, J. M., Prolactin and rapid eye movement, *Sleep,* 18, 536, 1995.
6. Heller, H. C. and Glotzbach, S. F., Thermoregulation during sleep and hibernation, *Int. Rev. Physiol.*, 15, 147, 1977.
7. Obál, Jr., F., Rubicsek, G., Alföldi, P., Sáry, G., and Obál, F., Changes in the brain and core temperatures in relation to the various arousal states in rats in the light and dark periods of the day, *Pflügers Arch.*, 404, 73, 1985.
8. Walter, J., Davenne, D., Shoham, S., Dinarello, C. A., and Krueger, J. M., Brain temperature changes coupled to sleep states persist during interleukin 1-enhanced sleep, *Am. J. Physiol.*, 250, R96, 1986.
9. Krueger, J. M., Walter, J., Dinarello, C. A., Wolff, S. M., and Chedid, L., Sleep-promoting effects of endogenous pyrogen (interleukin-1), *Am. J. Physiol.*, 246, R994, 1984.
10. Tobler, I., Borbély, A. A., Schwyzer, M., and Fontana, A., Interleukin-1 derived from astrocytes enhances slow-wave activity in sleep EEG of the rat, *Eur. J. Pharmacol.*, 104, 191, 1984.
11. Opp, M. R., Obál, Jr., F., and Krueger, J. M., Interleukin-1 alters rat sleep: temporal and dose-related effects, *Am. J. Physiol.*, 260, R52, 1991.
12. Susic, V. and Totic, S., "Recovery" function of sleep: effects of purified human interleukin-1 on the sleep and febrile response of cats, *Met. Brain Dis.*, 4, 73, 1989.
13. Fang, J., Kapás, L., Wang, Y., and Krueger, J. M., The TNF 55kD receptor and the IL-1 type I receptor are involved in physiological sleep regulation, *Soc. for Neurosci. Absts.*, 22, 147, 1996.
14. Friedman, E. M., Boinski, S., and Coe, C. L., Interleukin-1 induces sleep-like behavior and alters call structure in juvenile rhesus macaques, *Am. J. Primatol.*, 35, 145, 1995.
15. Dinarello, C. A., Interleukin-1 and interleukin-1 antagonism, *Blood*, 77, 1627, 1991.
16. Opp, M. R. and Krueger, J. M., Interleukin-1 is involved in responses to sleep deprivation in the rabbit, *Brain Res.*, 639, 57, 1994.

17. Opp, M. R. and Krueger, J. M., Anti-interleukin-1β reduces sleep and sleep rebound after sleep deprivation in rats, *Am. J. Physiol.*, 266, R688, 1994.
18. Takahashi, S., Kapás, L., Fang, J., Wang, Y., Seyer, J. M., and Krueger, J. M., An interleukin-1 receptor fragment inhibits spontaneous sleep and muramyl dipeptide-induced sleep in rabbits, *Am. J. Physiol.*, 271, R101, 1996.
19. Opp, M. R., Postlethwaite, A. E., Seyer, J. M., and Krueger, J. M., Interleukin 1 receptor antagonist blocks somnogenic and pyrogenic responses to an interleukin 1 fragment, *PNAS*, 89, 3726, 1992.
20. Opp, M. R., Smith, E. M., and Hughes, T. K., Interleukin-10 acts in the central nervous system of rats to reduce sleep, *J. Neuroimmunol.*, 60, 165, 1995.
21. Krueger, J. M., Kapás, L., Opp, M. R., and Obál, Jr., F., Prostaglandins E_2 and D_2 have little effect on rabbit sleep, *Physiol. and Behav.*, 51, 481, 1992.
22. Moldofsky, H., Lue, F. A., Eisen, J., Keystone, K., and Gorczynski, R. M., The relationship of interleukin-1 and immune functions to sleep in humans, *Psychosom. Med.*, 48, 309, 1986.
23. Gudewill, S., Pollmächer, T., Vedder, H., Schreiber, W., and Fassbender, K., Holsboer, F., Nocturnal plasma levels of cytokines in healthy men, *Eur. Arch. Psychiatry Clin. Neurosci.*, 242, 53, 1992.
24. Uthgenannt, D., Schoolman, D., Pietrowsky, R., Fehm, H. L., and Born, J., Effects of sleep on the production of cytokines in humans, *Psychosom. Med.*, 57, 97, 1995.
25. Hohagen, F., Timmer, J., Weyerbrock, A., Fritsch-Montero, R., Ganter, U., Krieger, S., Berger, M., and Bauer, J., Cytokine production during sleep and wakefulness and its relationship to cortisol in healthy humans, *Neuropsychobiology*, 28, 9, 1993.
26. Lue, F. A., Bail, M., Jephthah-Ocholo, J., Carayanniotis, K., Gorczynski, R., and Moldofsky, H., Sleep and cerebrospinal fluid interleukin-1 like activity in the cat, *Intern. J. Neurosci.*, 42, 179, 1988.
27. Taishi, P., Bredow, S., Guha-Thakurta, N., Obál, Jr., F., and Krueger, J. M., Diurnal variations of interleukin-1β mRNA and β-actin mRNA in rat brain, *J. Neuroimmunol.*, (in press).
28. Mackiewicz, M., Sollars, P. J., Ogilvie, M. D., and Pack, A. I., Modulation of IL-1β gene expression in the rat CNS during sleep deprivation, *NeuroReport,* 7, 529, 1996.
29. Plata-Salaman, C. R., Immunoregulators in the nervous system, *Neurosci. and Biobehav.*, 15, 185, 1991.
30. Miller, L. G., Galpern, W. G., Lumpkin, M., Chesley, S. F., Dinarello, C. A., Interleukin-1 augments γ-aminobutyric acid_A receptor function in brain, *Mol. Pharmacol.*, 39, 105, 1991.
31. Steriade, M. and McCarley, R. W., *Brainstem Control of Wakefulness and Sleep*, Plenum Press, New York, 1990, Chap. 1.
32. Auron, P. E., Quigley, G. J., Rosenwasser, C. J., and Gehrke, L., Multiple amino acid substitutions suggest a structural basis for the separation of biological activity and receptor binding in a mutant interleukin-1 beta protein, *Biochemistry,* 31, 6632, 1992.
33. Cunningham, Jr., E. T. and DeSouza, E. B., Interleukin-1 receptors in the brain and endocrine tissues, *Immunol. Today,* 14, 171, 1993.
34. Liu, C., Bai, Y., Ganea, D., and Hart, R., Species-specific activity of rat recombinant IL-1β, *J. Interferon and Cytokine Res.*, 15, 985, 1995.

35. Ericsson, A., Liu, C., Hart, R. P., and Sawchenko, P. E., Type I interleukin-1 receptor in the rat brain: distribution, regulation, and relationships to sites of IL-1-induced cellular activation, *J. Comp. Neurology,* 361, 681, 1995.
36. Colotta, F., Dower, S. K., Sims, J. E., and Mantovani, A., The type II 'decoy' receptor: a novel regulatory pathway for interleukin-1, *Immunol. Today,* 15, 562, 1994.
37. Greenfeder, S. A., Nunes, P., Knee, L., Labon, M., Chizzonite, R. A., and Ju, G., Molecular cloning and characterization of a second subunit of the interleukin-1 receptor complex, *J. Biol. Chem.*, 270, 13757, 1995.
38. Liu, C., Chalmers, D., Maki, R., and DeSouza, E. B., Rat homolog of mouse interleukin-1 receptor accessory protein: cloning, localization and modulation studies, *J. Neuroimmunol.*, 66, 41, 1996.
39. Wesche, H., Neumann, D., Resch, K., and Martin, M. U., Co-expression of mRNA type I and type II interleukin-1 receptors and the IL-1 receptor accessory protein correlates to IL-1 responsiveness, *FEBS Lett.*, 391, 104, 1996.
40. Cao, Z., Henzel, W. J., and Gao, X., IRAK: a kinase associated with the interleukin-1 receptor, *Science,* 271, 1128, 1996.
41. Lovenberg, T. W., Crowe, P. D., Liu, C., Chalmers, D. T., Liu, X. J., Liaw, C., Clevenger, W., Oltersdorf, T., DeSouza, E. B., and Maki, R. A., Cloning of a cDNA encoding a novel interleukin-1 receptor related protein (IL1R-rp), *J. Neuroimmunol.*, (in press).
42. Liu, C., Hart, R. P., Liu, X. J., Clevenger, W., Maki, R. A., and DeSouza, E. B., Cloning and characterization of an alternatively processed human type II interleukin-1 receptor mRNA, *J. Biol. Chem.*, (in press).
43. Krueger, J. M. and Obál, Jr., F., A neuronal group theory of sleep function, *J. Sleep Res.*, 2, 63, 1993.
44. Krueger, J. M., Obál, Jr., F., Kapás, L., and Fang, J., Brain organization and sleep function, *Behav. Brain Res.*, 69, 177, 1995.
45. Pigarev, I. N., Nothdurft, H. C., Rodionova, E. I., and Kastner, S., Asynchronous sleep development in cortical areas, *J. Sleep Res.,* 5 (Suppl. 1), 176, 1996.
46. Sambrook, J., Fritsch, E. F., and Mariatis, T., *Molecular Cloning*, Cold Spring Harbor Laboratory Press, Cold Spring Harbor, NY, 1989.
47. Ferre, F., Marchese, A., Pezzoli, P., Griffin, S., Buxton, E., and Boyer, V., Quantitative PCR: An overview, in *The Polymerase Chain Reaction*, Mullis, K. B., Fene, F., Gibs, R. A., Eds., Birkhauser, Boston, 1994, 67.
48. McPherson, M. J., Hames, B. D., and Taylor, G. R., PCR2, a practical approach, Oxford University Press, Oxford, England.

Chapter 5

Immediate Early Genes as a Tool to Understand the Regulation of the Sleep–Waking Cycle: Immunocytochemistry, *In Situ* Hybridization, and Antisense Approaches

Chiara Cirelli, Maria Pompeiano, and Giulio Tononi

Contents

0-8493-3361-X/98/$0.00+$.50

I. Introduction

Over the last few years, it has become clear that the activation or deactivation of the expression of specific genes can occur in a matter of hours or even minutes. This time frame is compatible with the duration of sleep–waking states and with the time constants of their regulation. Thus, it becomes relevant to ask whether gene expression in the brain changes across the sleep–waking cycle and after sleep deprivation.[1,2,9-11] Two different approaches can be used to study changes in gene expression between sleep and waking: in a *targeted* approach, one can examine the expression of a gene of interest (this chapter); in a *systematic* approach, one can examine changes in the expression of all mRNAs present in a given tissue (see Pompeiano et al., Chapter 13).

Our first attempt to detect changes in gene expression between sleep and waking focused on immediate early genes (IEGs) such as c-fos.[8] c-Fos encodes a transcription factor, Fos protein, that is induced by many extracellular stimuli. Typically, c-fos and other IEGs are the first genes to be turned on or off in the chain of events that leads to changes in the expression of other genes. Fos protein, for instance, has specific DNA-binding domains by which it can affect the expression of many target genes.[8] Recently, the use of antisense oligonucleotides targeted at Fos, as well as gene knockout approaches, have demonstrated that Fos can act as a transcription factor *in vivo* and produce functional and behavioral consequences.[5]

To study changes in the expression of Fos and other IEGs expression over the entire rat brain, a combination of radioactive *in situ* hybridization and immunocytochemistry is particularly useful. *In situ* mRNA imaging offers easy quantification and takes advantage of the fast induction times of mRNA, while protein imaging with immunocytochemistry provides cellular resolution and compatibility with other anatomical techniques. Through the combined use of these techniques we found that c-fos expression was increased in several brain areas, with respect to sleep, after a few hours of spontaneous waking.[1,9] These areas included the cerebral cortex, hippocampal formation, medial and lateral preoptic areas, and some thalamic and brainstem nuclei. A parallel series of studies[2,10] indicated that after a few hours of sleep deprivation the patterns of IEGs expression were remarkably similar to those observed after spontaneous wakefulness, suggesting that such patterns are associated with waking *per se*, rather than with circadian or stress factors. Recently we showed, using again both *in situ* hybridization and immunocytochemistry, that the expression of c-fos during waking is strictly dependent on the level of activity of the noradrenergic system.[6,13] An example of the high levels of c-fos during forced and spontaneous waking and of its low levels during sleep can be seen in Figure 5.1, which shows immunocytochemical staining of Fos protein applied to coronal sections of the cerebral cortex. Double labeling techniques can also be employed to demonstrate that cells expressing Fos protein during waking are neurons and not glial cells (Figure 5.2).

In our studies, the most consistent increase in Fos expression during waking was found in the preoptic area (POA) of the hypothalamus, a region that has previously been implicated in sleep regulation. Double labeling showed that Fos-positive cells activated by waking in the POA are not GABA-ergic.[4] Since the

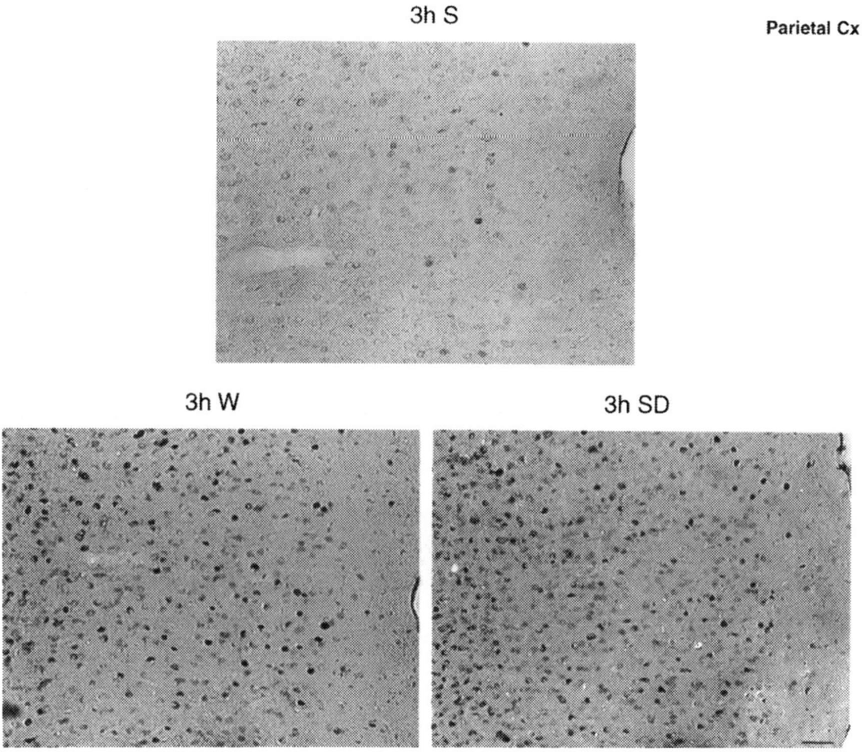

FIGURE 5.1
Changes in Fos protein levels in the parietal cortex of the rat during sleep and waking. The sleeping (S) rat was sacrificed after 3h of spontaneous sleep during the light hours. The waking (W) rat was spontaneously awake for 3h during the dark period. The sleep-deprived rat (SD) was kept awake for 3h during the light hours by gentle handling. Immunocytochemistry with an antibody against Fos (CRB, 1:2000) was performed on 40 μm free-floating coronal sections following the protocol given in the text. Scale bar (in lower right corner) is 10 μm.

increase in Fos protein expression in the POA was related, though not in a linear fashion, to the duration of prior waking,[2] the question arises whether Fos plays a causal role in the homeostatic control of sleep. This question can be addressed by using an approach based on the injection of oligonucleotides specific for c fos. With this technology, one targets Fos protein by using antisense oligonucleotides that interfere with the processing, transport, or translation of the corresponding mRNA. By using a c-fos antisense oligonucleotide injected locally in the POA, we specifically blocked Fos protein expression in the POA during waking. We found that the rats slept much less the day after the injection and there was no sleep rebound afterwards.[3] Thus, Fos protein expression in the POA during waking may be an integral part of the mechanisms that assess the duration and intensity of prior waking and/or of the homeostatic or executive mechanisms that bring about sleep.

The protocols that are given below refer to radioactive *in situ* hybridization for c-fos mRNA, immunocytochemistry for Fos protein, double-labeling immunocy-

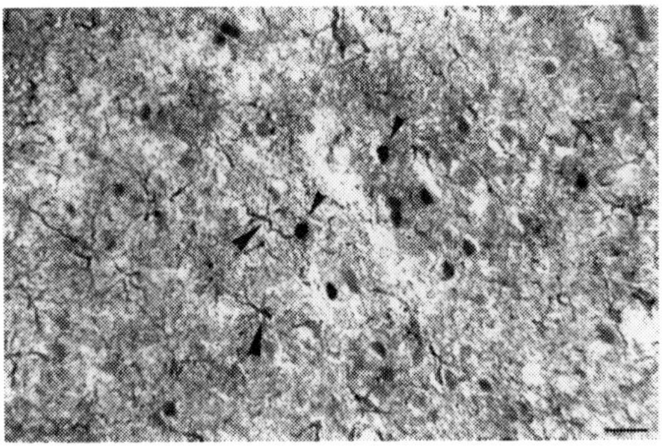

FIGURE 5.2
Double labeling in the parietal cortex of a sleep-deprived rat with antibodies against Fos and GFAP (glial fibrillary acidic protein) shows that Fos positive cells (small arrowheads) are not glial cells because they are GFAP negative. Large arrowheads indicate glial GFAP positive cells. Details of the protocol are given in the text. Scale bar is 20 μm.

tochemistry to assess the neuronal and/or glial nature of Fos-positive cells, and antisense injections using a c-fos oligonucleotide to interfere with Fos protein expression *in vivo*.

II. Protocols

A. Radioactive *In Situ* Hybridization with ^{32}P- or ^{33}P-Labeled Oligonucleotides Specific for c-Fos mRNA

1. Synthesis and Labeling of the Probe

The antisense oligonucleotide complementary to the base sequence coding for amino acid 1-16 of the Fos protein[7] is synthesized on a DNA synthesizer (391 Applied Biosystems, Foster City, CA) and purified on a 15% polyacrylamine/8 *M* urea preparative sequencing gel. The oligonucleotide (2 pmol) is tailed at its 3′ end using 25 units of the enzyme terminal deoxynucleotidyltransferase (Boehringer, Indianapolis, IN) and 16 pmol of $\alpha^{32}P$-dATP or $\alpha^{33}P$-dATP (3000 Ci/mmol; Du Pont, Boston, MA) in 100 m*M* sodium cacodylate, pH 7.2/2 m*M* $CoCl_2$/0.2 m*M* dithiothreitol to specific activities of 1 to 4×10^4 Ci/mmol. Labeled probes are purified by chromatography through a NACS PREPAC column (Gibco BRL, Gaithersburg, MD) according to the manufacturer's instructions.

2. Hybridization

Rats are rapidly anesthetized and killed by decapitation. Brains are quickly removed, frozen with powdered dry ice, and stored at –70°C until sectioned. Frontal sections (20 μm) are cut on a cryostat at –21°C and mounted on gelatin-coated slides. To minimize the variability due to incubation procedures, one section from each experimental group (sleep, sleep-deprivation, spontaneous waking) is mounted on the same slide. Slides are kept at –40°C until use.

1. Thaw out and air dry the sections at room temperature for about 20 min. Fixation and washes are done in 70 ml jars.
2. Fix sections in 4% (w/v) paraformaldehyde (cold) in 0.1 *M* phosphate buffer (pH 7.4) for 20 min at room temperature.
3. Wash 1 × 5 min in 3X PBS with gentle shaking at room temperature.
4. Wash 2 × 5 min in 1X PBS with gentle shaking at room temperature.
5. Incubate with predigested pronase for 1 to 10 min at room temperature with shaking. To prepare a stock solution of pronase: digest 1 g of pronase (Boehringer) in 10 ml distilled water for 4 h at 37°C; store in aliquots of 140 μl (for 1 jar of 70 ml) at –20°C. Immediately before use, dissolve 1 aliquot of pronase (140 μl) in 70 ml of pronase buffer (50 m*M* Tris-HCl, pH 7.5; 5 m*M* EDTA, pH 8.0).
6. Stop digestion with glycine (2 mg/ml in 1X PBS; wash for 1 min).
7. Wash 2 × 1 min in 1X PBS with gentle shaking at room temperature.
8. Dehydrate the sections through an ascending series of alcohols before starting the hybridization.
9. Hybridize the sections under a nescofilm coverslip overnight at 42°C. Hybridization buffer: 40% (v/v) formamide, 0.6 *M* NaCl, 1X Denhardt's, 10 m*M* Tris-HCl, pH 7.5, 1 m*M* EDTA, pH 8.0, 0.5 mg/ml tRNA, 20% (w/v) dextran sulfate. For 1 ml of hybridization buffer: 400 μl formamide, 120 μl 5 *M* NaCl, 20 μl 50X Denhardt's, 10 μl 1 *M* Tris-HCl (pH 7.5), 2 μl 0.5 *M* EDTA (pH 8.0), 48 μl tRNA (12 mg/ml yeast tRNA in distilled water), 200 μl dextran sulfate, 200 μl distilled water. The probe is used at 0.8 pmol/ml.
10. Wash 4 × 1 h at 55°C in washing buffer (0.6 *M* NaCl, 10 m*M* Tris, pH 7.5, 1 m*M* EDTA, pH 8.0).
11. Dehydrate and air dry the sections. Autoradiograms are generated by apposition to β-max film (Amersham, Arlington Heights, IL) for 20 (α^{32}P-dATP) or 50 (α^{33}P-dATP) days at –70°C. After exposure, the films are developed in Kodak D19.
12. To quantify the hybridization signals, film autoradiographs are digitized with a computer-based image analysis system. The optical density for each region for each subject is divided by the optical density of the white matter from the same section to obtain an optical density ratio (ODR). For each region, ODRs are collected in duplicate and on both sides of the brain. The nonparametric Mann-Whitney U-test can be used for the statistical analysis of the results.

The specificity of the hybridization signal can be verified by a series of control experiments.[12] In pilot experiments, two oligonucleotides complementary to different regions of c-fos mRNA should be used separately as hybridization probes in consecutive tissue sections, and should yield similar hybridization patterns. Specific hybridization signal should not be obtained when an excess (20X) of unlabeled oligonucleotide is included in the hybridization solution. A sharp decrease in intensity of the hybridization signal should be observed at a temperature consistent with the theoretical melting temperature of the hybrids. Pretreatment of the tissue sections with RNase should eliminate true hybridization signal, and hybridization with a sense probe should not show any positive signal. Finally, on total RNA from rat brain, the probe should identify a single band of the right size (2.2 kb).

B. Fos Immunocytochemistry

Under deep anesthesia rats are transcardially perfused with 0.9% (w/v) cold saline (50 ml) followed by 4% cold paraformaldehyde in 0.1 *M* phosphate buffer (350 ml; pH 7.4). Brains are removed, postfixed in 4% paraformaldehyde for 5 hours at 4°C, and cryoprotected in 20% w/v sucrose in 1X PBS. Freeze brains on dry ice and store them at –20 °C. Mount the brain on chuck with OCT compound (Miles, Elkhart, IN) in the cryostat set at –20°C. Cut the brain in frontal sections at 40 μm and place them into cold 1X PBS with 0.1% sodium azide. Store the sections at 4 °C.

1. Before incubating with primary antibody, wash the sections 1 × 5 min in cold 1X PBS. All procedures are performed in 50 ml beakers.
2. Incubate in 1X PBS containing the primary antibody (CRB [Cheshire, UK] sheep anti-Fos IgG, OA-11-824, 1:2000), 2% normal rabbit serum, and 0.3% Triton X-100 for 72 h at 4 °C on a shaker plate. For up to 10 sections, 2.5 ml of PBS are enough.
3. Wash 3 × 15 min in PBS 4% with shaking at room temperature.
4. Incubate 2 hours in 1X PBS containing secondary antibody (biotinylated anti-sheep IgG; 1:200; Vector, Burlingame, CA), 2% normal rabbit serum, 0.3% Triton-X-100 at room temperature on a shaker.
5. Wash 3 × 10 min in 1X PBS. Prepare ABC reagent (Elite kit, Vector) according to manufacturer's instructions.
6. Incubate for 1 h in ABC reagent (in 1X PBS containing 0.3% Triton X-100) at room temperature without shaking. Prepare ABC 0.5 to 1 h before use.
7. Wash 3 × 5 min in 1X PBS 4% with shaking at room temperature.
8. React in the chromogen DAB (Vector). In 2.5 ml distilled H_2O, add 1 drop of buffer stock solution and 2 drops of DAB stock solution. Mix and add your tissue. Mix and then add 1 drop of hydrogen peroxide solution and 1 drop of nickel solution. Keep mixing and stop the reaction when the staining is appropriate (1 to 10 min). Eventually check the reaction under the microscope.
9. Stop the reaction by filling the beaker with 1X PBS.
10. Wash sections 3 × 10 min in 1X PBS with shaking at room temperature.
11. If no further staining is desired, mount the sections from cold PBS onto subbed glass slides, dry on a slide warmer overnight. Sections are then dehydrated through an ascending series of alcohols, cleared in xylene, and coverslipped with Permount (see Figure 5.1).
12. The number of Fos positive cells can be counted on the basis of camera lucida drawings, or, in the case of areas with a large number of positive cells, with a computer-assisted imaging system interfaced with the microscope. Computer-generated outlines of the regions of interest are superimposed on the sections and the number of stained cells is counted. The areas of the counting regions are determined by the computer, and cell

densities are calculated. The background level is set such that only cells with unequivocally positive, darkly stained nuclei are counted. The nonparametric Mann-Whitney U-test can be used for the statistical analysis of the results.

Immunocytochemistry for Fos protein can also be carried out on slide-mounted sections from animals that were not perfused. In this case, sections are postfixed as described in the protocol for *in situ* hybridization. The quality of the tissue from frozen, post-fixed sections is not as good as with free-floating sections from perfused animals, but the intensity of Fos staining and the level of background are comparable (compare with Figures 5.1 and 5.3). The great advantage of this methodology is that both *in situ* hybridization and immunocytochemistry can be performed on tissue from the same animal. Conversely, while it is theoretically possible to perform *in situ* hybridization on perfused tissue, we have never succeeded in obtaining a good signal.

C. Double Labeling Using Fos, Neuronal, and Glial Markers

In order to identify the type of cells (neurons or glia) that express Fos or other immediate early genes, a double labeling technique is used with an anti-Fos antibody and with anti-glial fibrillary acidic protein (GFAP) and microtubule-associated protein 2 (MAP 2) antibodies.

1. Complete steps 1 through 10 as above but decrease the dilution of the primary anti-Fos antibody to 1:1000.
2. Incubate the sections in 1X PBS containing anti-glial fibrillary acidic protein antibody (GFAP, Sigma, St. Louis, MO, 1:400) or antimicrotubule-associated protein 2 antibody (MAP-2, Sigma, 1:250), 2% normal horse serum and 0.1% Triton-X-100 overnight at room temperature on a shaker plate.
3. Wash 3 × 15 min in PBS with shaking at room temperature.
4. Incubate 2 h in PBS containing secondary antibody (biotinylated anti-mouse IgG; 1:200; Vector), 2% normal horse serum, 0.1% Triton-X-100 at room temperature on a shaker.
5. Complete steps 5 through 11 as above but do not add nickel solution to the chromogen solution. In this way the dark (black) nuclear staining for Fos can be distinguished from the brown cytoplasmic staining for GFAP and MAP-2 (see Figure 5.2).

D. *In Vivo* Use of a c-Fos Phosphothioate Antisense Oligonucleotide

The protocol for antisense injections is closely tied to the particular experimental question that is being addressed. For this reason, we describe below the procedures used in a series of experiments in which we examined whether Fos expression in

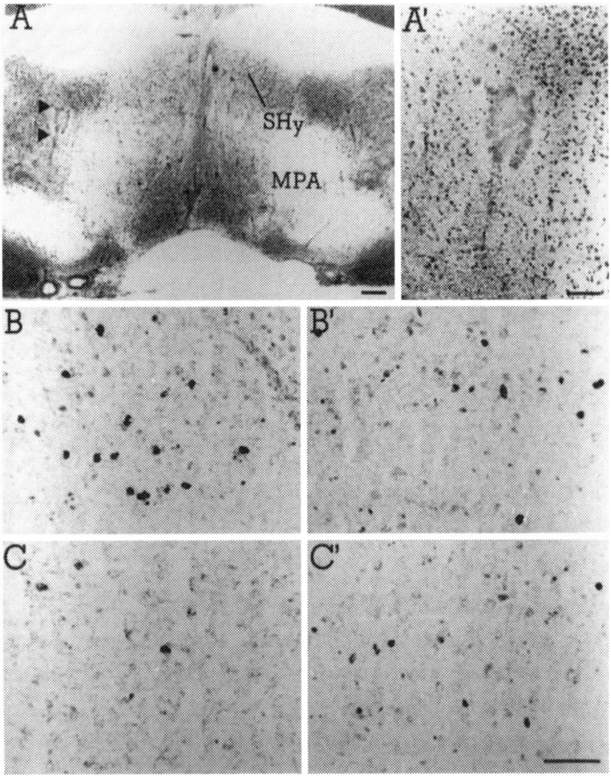

FIGURE 5.3
Blocking of Fos expression during waking by local injection of a c-fos antisense oligonucleotide in the medial preoptic area of the rat. (A) Photomicrograph of the injection area stained with cresyl violet. The arrowheads indicate the track of the cannula. (A′) Higher magnification of the track of the cannula. (B, C) Fos positive cells in the septohypothalamic nucleus and medial preoptic area, respectively, ipsilateral to the injection side. (B′, C′) Fos positive cells are present in the septohypothalamic nucleus and medial preoptic area, respectively, contralateral to the injection side. Note that the Fos induction that is usually seen during waking in the medial preoptic area is blocked on the side of the injection of an antisense oligonucleotide against c-fos. Details of the protocol are given in the text. Scale bars are 100 μm in A, and 50 μm in A′ and C′.

the POA plays a causal role in sleep regulation.[3] In these studies, rats were implanted with electrodes for electrocorticography (ECG) and electromyography (EMG) and equipped with two 24G stainless steel guide cannulae aimed at the transition between the medial and the lateral preoptic area (A –0.5; L 1.0; H 7.0). They were polygraphically recorded until the percentages of sleep and waking were regular (LD 12:12, light on at 8 A.M.).

c-Fos antisense (5′–GAACATCATGGTCGT–3′, centered on the first AUG of the rat mRNA) and sense (5′–ACGACCATGATGTTC–3′) sequences were synthesized as phosphothioate modified oligodeoxynucleotides on a DNA synthesizer (391 Applied Biosystems) and purified on a 15% polyacrylamide/8 *M* urea sequencing gel.

Injections of antisense oligonucleotides were performed over a period of 2 min using a 30G needle connected to a 1μl Hamilton syringe. Two nmol of antisense oligonucleotide (in 1 μl mock-CSF) were effective when injected locally, while 1 nmol produced a mild suppression of Fos expression, and 0.5 nmol were totally ineffective.[5] The needle was retracted 2 min or more after the injection. The injections were performed unilaterally as well as bilaterally, and no more than 1 to 2 injections were performed on each side because tissue damage with reactive gliosis has been reported after 3 to 4 local infusions on a daily basis.[5]

After either 11 or 36 hours the animals were sacrificed, the brains were quickly removed, frozen on dry ice, and frontal sections (20 μm) were cut on a cryostat. A few frontal sections of the brain were cut and stained with cresyl violet for histological examination of the injection site to assess tissue damage. On the other sections, immunocytochemistry for Fos protein was performed as described previously.

Several technical points should be mentioned about the use of antisense injections *in vivo*, as exemplified by these experiments.

During the first hours after the injection, it is common to observe aspecific behavioral effects due to irritative phenomena.[5] These effects are aspecific because they are produced both by antisense injections and by sense injections (see below), or even by the vehicle alone.

It takes a few (4 to 6) hours for the oligonucleotide to be taken up by the cells and to reach the nucleus. In most cases, the specific effects last for 10 to 15 hours after the injections, and any residual effect disappears within 24 hours.[5] Thus, in our studies, to assess the effects of antisense injections on the sleep–waking cycle, rats were continuously recorded the day before the injection, the day of the injection, and the day after the injection.

Before attempting to interpret the results, it is essential to check whether the antisense oligonucleotide is able to block the expected induction of Fos expression. In our study, some animals received only a unilateral injection and were sacrificed 11 h after the injection, at around 2 A.M. (i.e., after 6 h of spontaneous waking). We found that on the side of the injection, no or very few Fos positive cells were present in the POA, while they were abundant on the contralateral side.

Control experiments to ascertain whether the effects of c-fos antisense oligonucleotides are specific are essential. They include the use of sense and/or missense oligonucleotides, injected at the same dose as the antisense oligonucleotides. Sense oligonucleotides have a sequence complementary to that of the antisense, while missense oligonucleotides have a random sequence but with the same CG/AT ratio as the antisense oligonucleotide, thus retaining the same binding energy to a complementary sequence. Sense and/or missense oligonucleotide injections should not interfere with Fos expression on the injection side and should not produce the effects on the sleep–waking cycle observed with the antisense oligonucleotide. One should also control whether a c-fos antisense oligonucleotide does interfere with the expression of other IEGs. Another way to control for specificity is to show that the reduction in Fos expression observed after a c-fos antisense oligonucleotide injection is associated with a reduction of AP-1 binding activity.[5]

III. Discussion

The three techniques presented here to study the expression of immediate early genes across behavioral states are complementary. Because of the fast induction time of mRNA (a few minutes), *in situ* hybridization is the preferred method of use to document very rapid changes in gene expression that can occur during the transition between one behavioral state and another. In addition, when performed with radioactive probes, *in situ* hybridization is easy to quantitate and very sensitive, allowing the detection of small differences in the expression of genes between sleep and waking. Immunocytochemistry, on the other hand, provides cellular resolution and compatibility with other anatomical techniques. This advantage is particularly evident when studying Fos protein expression. The nuclear staining for Fos protein can be easily combined in double-labeling experiments with cytoplasmic staining using other antibodies or tracing methods. The combined use of immunocytochemistry and *in situ* hybridization is also useful because in some brain regions (e.g., the cerebellum) changes in mRNA levels may not be followed by changes in protein levels.

A role for c-fos as a marker of "genetic" activation can only be suggested if its protein product, Fos, that can act as a transcription factor, is induced. Even then, whether Fos is responsible for changes in physiological functions and behavior needs to be assessed directly in each experimental condition. Antisense techniques have already proven useful in this respect in several different experimental paradigms. They have been particularly successful when applied to target Fos expression in the brain because: 1) neurons are less equipped than other cell types with enzymes that degrade oligonucleotides, 2) the "basal" level of expression of c-fos is often very low in neurons, a property that makes neurons more sensitive to the action of antisense oligonucleotides, and 3) IEGs such as c-fos are strategically placed in the pathway linking extracellular signals to gene transcription. The antisense technique has the important advantage of being both very specific and relatively simple. However, careful consideration should be given to factors such as the stability of modified oligonucleotides, the timing and duration of the injections, and the levels of expression of the targeted mRNAs.[5] As more becomes known about specific gene products, the use of antisense approaches should become a standard tool to causally interfere with the expression of a specific gene in a specific brain region at a specific time.

References

1. Cirelli, C., Pompeiano, M., and Tononi, G., Fos-like immunoreactivity in the rat brain in spontaneous wakefulness and sleep, *Arch. Ital. Biol.*, 131, 327, 1993.
2. Cirelli, C., Pompeiano, M., and Tononi, G., Sleep deprivation and c-fos expression in the rat brain, *J. Sleep Res.*, 4, 92, 1995.
3. Cirelli, C., Pompeiano, M., Arrighi P., and Tononi, G., Sleep–waking changes after c-fos antisense injections in the medial preoptic area, *Neuroreport*, 6, 801, 1995.
4. Cirelli, C., Pompeiano, M., Arrighi P., and Tononi, G., Fos-positive cells associated with forced wakefulness in the hypothalamus of the rat are not GABAergic, *Arch. Ital. Biol.*, 133, 143, 1995.

5. Cirelli, C., Pompeiano, M., and Tononi, G., *In vivo* antisense approaches to the role of immediate early gene expression in the brain, *Regulatory Peptides*, 59, 151, 1995.
6. Cirelli, C., Pompeiano, M., and Tononi, G., Neuronal gene expression in the waking state: a role for the locus coeruleus. *Science*, 274, 1211, 1996.
7. Curran, T., Gordon, M.B., Rubino, K.L., and Sambucetti, L.C., Isolation and characterization of the c-fos(rat) cDNA and analysis of posttranslational modification *in vitro*. *Oncogene*, 2, 79, 1987.
8. Hughes, P. and Dragunow, M., Induction of immediate-early genes and the control of neurotransmitter-regulated gene expression within the nervous system, *Pharmacol. Rev.*, 47, 133, 1995.
9. Pompeiano, M., Cirelli, C., and Tononi, G., Immediate-early genes in spontaneous wakefulness and sleep: Expression of c-fos and NGFI-A mRNA and protein, *J. Sleep Res.*, 3, 80, 1994.
10. Pompeiano, M., Cirelli, C., and Tononi, G., Effects of sleep deprivation on fos-like immunoreactivity in the rat brain, *Arch. Ital. Biol.*, 130, 325, 1992.
11. Pompeiano, M., Cirelli, C., Ronca-Testoni, S., and Tononi, G., NGFI-A expression in the rat brain after sleep deprivation. *Mol. Brain Res.*, 46, 143, 1997.
12. Tecott, L.H., Eberwine, J.H., Barchas, J.D., and Valentino, K.L. Methodological consideration in the utilization of in situ hybridization, in *In Situ Hybridization. Applications to Neurobiology,* K.L. Valentino and J.D. Barchas, Eds., Oxford University Press, New York, 1987, 3.
13. Tononi, G., Cirelli, C., and Pompeiano, M., Changes in gene expression during the sleep–waking cycle: a new view of activating systems, *Arch. Ital. Biol.*, 134, 21, 1995.

Chapter 6

Methods for the Measurement of Adenylyl Cyclase Activity

Charles W. Emala

Contents

I. Introduction

Adenylyl cyclase (AC) is a family of enzymes that synthesize cAMP. At least nine isoforms of adenylyl cyclase have been cloned and expressed[1-10] (and unpublished mouse type IX Gen Bank accession noU30602) which show unique tissue distributions and regulatory patterns. Molecules that have differential effects on individual

0-8493-3361-X/98/$0.00+$.50

adenylyl cyclase isoforms include Ca^{2+}/calmodulin,[11] G protein α subunits,[11] G protein βγ subunits,[12] protein kinase A,[13] and protein kinase C.[14,15] All of the known adenylyl cyclase isoforms are expressed in neural tissue, and some isoforms express unique distributions within the central nervous system.[11] The distributions of mRNA encoding specific adenylyl cyclase isoforms within different brain regions has been extensively characterized by *in situ* hybridization. Many cells express multiple isoforms in varying abundance which likely affords the cell an integrated control over net cAMP levels. Attempts to identify individual adenylyl cyclase proteins within a tissue or cell is hampered by the lack of specific antibodies and by the relatively low level of protein expression (0.01 to 0.001% of membrane protein[11]) of these enzymes. Additionally, attempts to selectively activate or inhibit only certain isoforms is hampered by the lack of specific effectors. Despite these limitations, enormous insight into the pivotal role that cAMP plays in many critical cell functions has been gained by the measurement of total adenylyl cyclase activity.

Many different receptors in neural tissues couple to the stimulation (opioids,[16] serotonin,[17] norepinephrine,[17] dopamine[18]) or inhibition (cannabinoids,[19] somatostatin,[20] muscarinic,[21] gamma-aminobutyric acid (GABA),[22] dopamine[18]) of adenylyl cyclase. Measurements of adenylyl cyclase activity in neural tissues have traditionally used broken cell preparations which allow selective activation of individual components of the receptor-G protein-adenylyl cyclase cascade. Activation or inhibition of adenylyl cyclase by a given receptor is mediated through an intermediary G protein. For some applications it may be desirable to bypass the receptor and study effectors which act at the level of the G protein or the adenylyl cyclase enzyme itself. In broken cell preparations G proteins can be directly activated by effectors such as GTP or AlF_3, and the adenylyl cyclase enzyme can be directly stimulated by forskolin or $MnCl_2$ (although the selectivity of forskolin for directly activating adenylyl cyclase independent of $G_s\alpha$ is not absolute since forskolin-stimulated adenylyl cyclase activity is further enhanced by the presence of $G_s\alpha$[23,24]). The ability to stimulate the pathway at progressively more distal sites forms a strategy to localize where in the receptor-G protein-adenylyl cyclase pathway that a change in function may be occurring.

The protocol described in this chapter will assume that a cell or tissue crude homogenate is being used as the "plasma membrane" fraction. Many modifications of the adenylyl cyclase assay have been described which include the substitution of acutely lysed whole cells in culture for cell plasma membrane fractions,[25] the quantitation of cAMP synthesized by means of a commercially available RIA to obviate the need for column chromatography,[26] or the preloading of intact cultured cells[27,28] or brain slices[29] with ^3H-adenine to form ^3H-ATP as a substrate for the adenylyl cyclase enzyme. The protocol described measures the ability of adenylyl cyclase enzymes within a crude tissue homogenate to synthesize ^{32}P-cAMP from ^{32}P-α-ATP. In addition to the trace amounts of ^{32}P-α-ATP substrate, the reaction mixture contains a buffering agent (Hepes or Tris, pH 7.4 to 8.0) with Mg^{2+} (a necessary cofactor for G protein activation), a chelator (EDTA or EGTA) to define the final free Mg^{2+} concentration and to remove trace divalent cations that are protease cofactors, unlabeled ATP (necessary for adenylyl cyclase activity not to be substrate limited), unlabeled cAMP (to serve as a preferred substrate for phosphodiesterases present in the membrane preparation), bovine serum albumin (to serve as a preferred substrate for proteases), and an ATP regenerating system. Since most membrane preparations include phosphatase enzymes that degrade ATP to ADP and ADP to AMP, an enzyme regeneration system is included to regenerate the pool of ATP substrate. Phosphocreatine is commonly used as a phosphate donor along with phosphocreatine kinase to regenerate ATP from ADP. In some systems it may be useful to also include a regenerating system to form ADP from AMP using myokinase (125 units/ml). Some investigators also add a nonselective phosphodiesterase inhibitor (e.g., isobutylmethylxanthine (IBMX)), to the Rodbell buffer, to ensure complete inhibition of phosphosdiesterases. We have found that the inclusion of unlabeled cAMP serves as a preferred substrate for phosphodiesterases, and the further addition of a phosphodiesterase inhibitor does not influence the recovery of ^{32}P-cAMP.

Adenylyl cyclase activity is commonly measured under basal (unstimulated) conditions and in response to activators effective at various levels of the receptor-G protein-adenylyl cyclase cascade. The receptor agonist to be studied is cell specific and will depend upon the specific question of interest. Other commonly used effectors include those which directly stimulate G proteins (GTP, nonhydrolyzable GTP analogs (GTPγS or GppNHp) and AlF_3) and effectors which directly activate the enzyme adenylyl cyclase (forskolin and Mn).

Following an incubation period of 5 to 30 minutes, the adenylyl cyclase reaction is stopped by the addition of a buffer that includes a detergent (SDS) to solubilize the membrane. This stop buffer also includes a buffering agent, unlabeled cAMP (to again serve as a preferential substrate for contaminating phosphodiesterases) and a known quantity of ^{3}H-cAMP. (By adding a known quantity of ^{3}H-cAMP to each sample and measuring the column recovery of ^{3}H-cAMP, individual column efficiencies for the recovery of cAMP can be calculated.) The reaction mixture is diluted and subjected to column chromatography[30,31] as described below.

II. Protocol

A. *Assembly of Dowex and Alumina Columns*

Sequential column chromatography over dowex and alumina resins is a procedure for the retention of unmetabolized ^{32}P-α-ATP within the resins and the elution of newly synthesized ^{32}P-cAMP into scintillation vials to measure incorporated ^{32}P.[30] The addition of a known amount of ^{3}H-cAMP to each reaction as a component of the stop buffer allows the calculation of the efficiency of each column's recovery of cAMP. The ^{32}P-count measured in each sample is then corrected for the individual column recovery to provide a more accurate determination of the amount of ^{32}P-cAMP synthesized in each tube.

The assembly of dowex and alumina columns in racks that allow stacking of the dowex over the alumina and the collection of column eluates into scintillation vials is the investment that has discouraged most investigators from establishing this assay in their laboratory. However, the commercial availability of stackable column racks (Kontes, Vineland, NJ, cat. no. 420201-0000) with disposable columns (Kontes cat. no. 420160-0000) and the fact that columns can be regenerated between assays and used for 30 to 60 assays before needing replacement has made the development of this assay much easier. A column containing dowex and a column containing alumina will be needed for each sample. The plastic disposable columns used are 20 cm in length, have an internal diameter of 8 mm, and include a funnel top and outlet cap that contains a bed support that holds the column resins within the column. This column easily accommodates 10-ml volumes that will be used to regenerate the resins (dowex and alumina) between each assay.

1. To prepare the dowex columns, the plastic components of the disposable columns are assembled and placed upright within the column racks.
2. A slurry of dowex resin (AG 50W-X4, 200 to 400 mesh, hydrogen form, BioRad) is prepared in distilled/deionized water (dH_2O) and while slowly mixing, the slurry is withdrawn by pipette and deposited into the column until the wet dowex settles to a height of 4 cm.
3. The alumina (activity grade, super I, type WN-6 neutral, Sigma) columns are prepared in separate plastic disposable columns. The alumina can be added to the columns as dry powder until the alumina resin bed achieves a height of 3 cm in the bottom of the column.
4. The columns can now be washed to prepare them for the first assay. This procedure will be repeated between each assay to regenerate the columns and can be repeated indefinitely until the columns' recoveries fall below 75% (monitored by the recovery of known quantities of ^{3}H-cAMP spiked into each sample tube as part of the stop buffer), at which time the resins must be replaced. The dowex columns are regenerated by one 10-ml wash with 1 *N* HCl followed by three 10-ml washes with dH_2O. The alumina columns are regenerated with one 10-ml wash with dH_2O and one 10-ml wash with 0.1 *M* imidazole, pH 7.5.

B. Preparation of Cellular Homogenates or Plasma Membrane Fractions for Adenylyl Cyclase Assay

A wide variety of broken cell preparations have been used for the measurement of adenylyl cyclase activity. The amount of starting tissue or cultured cells required to perform the assay can vary greatly depending upon the endogenous adenylyl cyclase activity, the care taken during membrane preparation, the degree of purification of the plasma membrane fraction, and the effectors that will be used to stimulate or inhibit adenylyl cyclase activity during the assay. Typically 5 to 50 μg of membrane protein will be used in each assay tube which is equivalent to 50 to 500 μg of tissue wet weight. Membranes should be prepared from tissues or cells as soon as they are harvested from the animal or cell culture dish to minimize loss of adenylyl cyclase activity. Because the adenylyl cyclase enzyme is a plasma membrane-associated protein, any cellular preparation that includes the plasma membrane should be suitable for the measurement of AC activity. Greater degrees of cell fractionation and plasma membrane purification may be necessary in some tissues to achieve measurable levels of AC activity. Cell or tissue homogenization can be achieved by a wide variety of methods, but every effort should be made to keep the preparation cold (4°C) during homogenization to preserve the activity of the adenylyl cyclase enzyme. A reasonable starting point is to prepare a cellular homogenate that is enriched for the plasma membrane fraction by first lysing the tissue or cultured cells by high-speed cutting blades, Potter-Elvehjem tissue grinder, or nitrogen cavitation. This homogenate can then be subjected to low-speed centrifugation (400 × *g*, 15 minutes, 4°C) to remove intact cells and large tissue and cellular debris. The supernatant containing the disrupted plasma membranes (and other cellular components) is then subjected to high-speed centrifugation (50,000 × *g*, 30 minutes, 4°C) to concentrate the cellular homogenate. This pellet can be resuspended in a suitable buffer and, if desired, washed several times to remove soluble cellular components before resuspending the final pellet at a protein concentration of 1 to 6 mg/ml. This cellular homogenate can then be used immediately for adenylyl cyclase assays or stored as aliquots at –70°C for up to 18 months for future determinations of adenylyl cyclase activities.

1. Obtain 0.5 to 5.0 grams of finely minced tissue, or several confluent T-175 flasks of cultured cells. Suspend the tissue/cells in 5 to 10 ml cold (4°C) homogenate buffer (e.g., 100 m*M* Hepes, pH 7.4, 1 m*M* EDTA) *Hint: in tissues that contain high protease activity some investigators have found it useful to add a cocktail of protease inhibitors to the buffer (e.g., 25 μg/ml leupeptin and/or 25 μg/ml aprotinin)* in a vessel suitable for homogenization with high-speed cutting blades which can then be combined with a Potter-Elvehjem tissue grinder with a motor driven (hand drill) pestle to optimize cell disruption.
2. Transfer the cell/tissue homogenate to a prechilled centrifuge tube and centrifuge at 400 × *g* for 15 minutes at 4°C. Transfer the supernatant to a clean prechilled centrifuge tube and centrifuge at 50,000 × *g*, 30 minutes at 4°C. Discard the supernatant and wash the pellet twice by resuspending it in a small volume (2 to 3 ml) of cold homogenate buffer and repeat the 50,000 × *g* centrifugation for 30 minutes at 4°C. The tissue/cellular homogenate can then be used immediately or frozen at –70°C.

C. *Assembly of Adenylyl Cyclase Assay*

1. The adenylyl cyclase incubation is performed in triplicate in a final volume of 100 μl in 10 × 75 mm borosilicate glass tubes in a 30°C water bath for 5 to 30 minutes. For convenience, all tubes first receive 50 μl of 2X Rodbell buffer, containing ^{32}P-α-ATP and an ATP regeneration system (creatine phosphate and creatine phosphokinase) (see Reagents Needed). The second 50 μl includes the effector to be tested (10 μl of 10X stock added to final reaction volume of 100 μl), the membrane preparation (which should be added last and is considered to be the start of the reaction) and water to bring the final volumes to 100 μl. It is most convenient to assemble the reaction in an ice bath and, following the addition of membranes, briefly vortex and transfer the tubes to a 30°C water bath.
2. Following the desired time of incubation (5 to 30 minutes) the reaction is terminated by the addition of 100 μl of stop buffer (see Reagents Needed), and the tubes are boiled in a heating block for 3 minutes to ensure that all adenylyl cyclase activity is inactivated.

3. The reaction mixtures (200 μl) can now be frozen (–20°C) for several days or carried immediately to the column chromatography protocol.
4. Three types of standards are prepared during the assembly and performance of the adenylyl cyclase assay. The first three tubes of the assay are water blanks or no membrane controls that are incubated along with all assay tubes and contain 2X Rodbell buffer (containing ^{32}P-α-ATP) and receive stop buffer (containing ^{3}H-cAMP). These tubes will serve to control for background ^{32}P counts that elute from the columns but do not represent newly synthesized ^{32}P-cAMP (since these tubes contain no membranes). The second standard for each assay is the ^{32}P standard. This is simply a measure of the initial ^{32}P-α-ATP supplied to each tube. It is most convenient (and accurate) to add 50 μl of 2X Rodbell buffer to a scintillation vial while dispensing the 2X Rodbell buffer into the assay tubes. The third standard is a ^{3}H-cAMP standard which is a measure of the amount of ^{3}H-cAMP counts added to each tube with the stop buffer. It is most convenient to dispense 100 μl of stop buffer into a scintillation vial while the stop buffer is being added to each reaction tube. A protocol sheet for a typical adenylyl cyclase reaction is shown in Table 6.1.

D. Column Chromatography to Separate Newly Synthesized ^{32}P-cAMP from ^{32}P-α-ATP Substrate

1. Each sample tube now contains 200 μl volume (100 μl of the original reaction volume and 100 μl of the stop buffer). To optimize recovery of this volume from the tube, an additional 800 μl of dH_2O is added to the tube and its contents dumped into the funnel attachment on top of a regenerated dowex column. The volume is allowed to drain by gravity. Each of the successive wash and elution volumes should be allowed to drain by gravity before adding the next wash.
2. The sample is now washed into the dowex columns by two 1 ml washes with dH_2O which are discarded.
3. When the columns finish dripping, the box of dowex columns is mounted above the box containing the regenerated alumina columns so that the eluate of each dowex column will drain into the funnel top of its corresponding alumina column.
4. Two 2-ml washes of H_2O are now applied to the dowex columns so that the water passes sequentially through the dowex and then alumina. These washes are discarded.
5. The dowex over alumina stacked columns are now mounted over a rack containing 21-ml glass scintillation vials to collect the columns' eluates.
6. One ml of dH_2O is now applied to the dowex column and allowed to drip into and through the alumina and is collected in its corresponding scintillation vial.
7. The dowex columns are then removed from the alumina, and 5 ml of 0.1 *M* imidazole, pH 7.5 is applied to the alumina columns and collected in the same scintillation vials. This 6-ml aqueous sample within the scintillation vials now contains the eluted ^{32}P-cAMP (generated from ^{32}P-α-ATP during the assay) and ^{3}H-cAMP (a known amount added to each sample tube as part of the stop buffer for measurement of individual column recoveries).
8. Sufficient scintillation cocktail is then added to each vial to completely solubilize the aqueous sample. 14 ml of EcoLite (ICN, Costa Mesa, CA) will solubilize 6 ml of aqueous sample and prevent separation of the organic and aqueous phases to allow accurate scintillation counting.

TABLE 6.1
Adenylyl Cyclase Assay Sample Protocol (All Volumes in μl)

Tube no.	2X hot Rodbell	Membrane	H_2O	10X Effector Stock Concentrations: GTP 100 μ*M*	Isoproterenol 10 μ*M*	NaF 0.1 *M*	$AlCl_3$ 1 m*M*	Forskolin 100 μ*M*
1–3[a]	50	0	50	0	0	0	0	0
4–6[b]	50	10 (20 μg)	40	0	0	0	0	0
7–9[c]	50	10	30	10	0	0	0	0
10–12[d]	50	10	20	10	10	0	0	0
13–15[e]	50	10	20	0	0	10	10	0
16–18[f]	50	10	30	0	0	0	0	10

[a] Tubes 1 through 3 receive all buffer components but no membranes. The ^{32}P counts recovered from these tubes will constitute ^{32}P-α-ATP that is not trapped on the columns plus background counts. Thus these tubes serve as "water blanks," and the counts generated from the tubes will be subtracted from all other values.

[b] Tubes 4 through 6 receive all buffer components plus membranes. ^{32}P counts recovered from these tubes above those counts recovered from tubes 1 through 3 will represent the basal or unstimulated adenylyl cyclase activity in the membrane preparation.

[c] Tubes 7 through 9 will measure the effects of GTP activation of all G proteins. Because all hormones that act through a receptor require some availability of GTP to allow receptor-G protein coupling and activation, GTP is added so as to not be of limiting quantity in the reaction when receptors such as β-adrenergic receptors (tubes 10 through 12) are activated. Because GTP itself will have some effects (stimulatory or inhibitory) on measured adenylyl cyclase activity, the GTP activity will be compared to the activity measured in the presence of GTP plus hormone (e.g., isoproterenol, tubes 10 through 12).

[d] Tubes 10 through 12 will measure the effect of a hormone stimulus on adenylyl cyclase activity. The example used here is isoproterenol stimulation of β-adrenergic receptors, but many different hormones are known to have either stimulatory or inhibitory effects, and it is the selection of this hormone that is most variable between studies and investigators.

[e] Tubes 13 through 15 will measure the effect of direct activation of all G proteins with Al/Cl_3. Many investigators add NaF to see this effect relying on contaminating Al^{3+} to form the true stimulatory ion AlF_3. We prefer to add $AlCl_3$ to ensure that Al^{3+} is not limiting or variable between tissue samples.

[f] Tubes 16 through 18 will measure the effect of direct activation adenylyl cyclase by adenylyl cyclase. Although forskolin directly activates adenylyl cyclase, the activity achieved has been shown to be influenced by the addition of G protein α subunits.[23] If more selective direct activation of the adenylyl cyclase enzyme without the influence of G protein α subunits is desired, Mn may be a more preferable substitute.[24]

E. Calculation of pmoles of cAMP Generated in Each Tube

The samples and standards are counted in a scintillation counter programmed to simultaneously count ^3H and ^{32}P. The β particle energies emitted by the radioactive decay of these isotopes are sufficiently separated that the counts can be distinguished on the basis of the intensities of the light pulses produced. There is a small overlap or spillover of the ^{32}P window of intensities into the ^3H window, and this can be calculated by the use of the ^{32}P standards that contain only ^{32}P. Channels 0 to 400 represent ^3H decay and channels 400 to 1000 represent ^{32}P decay in a Beckman LS 5000TD scintillation counter.

The amount of ^{32}P-cAMP generated is calculated from the number of ^{32}P counts in each sample. However, these counts need to first be corrected in four ways: (1) The number of background counts needs to be subtracted from all counts in both the ^{3}H window and ^{32}P window, (2) the spillover of ^{32}P counts into the ^{3}H window needs to be calculated and added back to each ^{32}P count, (3) the efficiency of column recovery needs to be calculated for each sample and the corresponding ^{32}P count increased to account for less than 100% column recovery, and (4) the amount of corrected ^{32}P counts from the tubes with no membranes (accounting for ^{32}P-α-ATP not trapped on the columns) needs to be subtracted from each sample. The corrected ^{32}P counts of each sample can then be used to calculate the amount of cAMP synthesized in pmoles/mg membrane protein/time. In our laboratory these arithmetic computations are carried out using a computer program written in GW Basic.

An example of the counts obtained for the sample assay given above is presented in Table 6.2. Using these sample counts the 4 steps necessary to calculate the

TABLE 6.2
Sample ^{3}H and ^{32}P Counts (cpm) Obtained from the Adenylyl Cyclase Assay Described in Table 6.1

	I. ^{3}H counts (channel 0–400)	II. ^{32}P counts (channel 400–1000)	III. % column recoveries
A. Background blank	45	60	
B. ^{32}P standard (1/20 of standard counted)	2402	80128	
C. ^{3}H standard	52673	62	
D. No membrane control sample tubes 1–3	42719	98	0.81
	40178	89	0.76
	44123	107	0.84
E. Basal activity sample tubes 4–6	43673	588	0.83
	39125	434	0.74
	45420	576	0.86
F. GTP-stimulated sample tubes 7–9	44357	829	0.84
	42367	799	0.80
	45676	839	0.87
G. GTP + Isoproterenol sample tubes 10–12	41126	1083	0.78
	45484	1344	0.86
	46252	1209	0.88
H. NaF/AlCl$_3$ sample tubes 13–15	45629	3098	0.87
	40354	2499	0.77
	42010	2609	0.80
I. Forskolin sample tubes 16–18	42099	2121	0.80
	39898	2033	0.76
	37656	1879	0.71

corrected ^{32}P counts will be presented. (1) Subtract background counts (A) of both isotopes from all other values (i.e., subtract 45 from all ^{3}H counts and subtract 60 from all ^{32}P counts). (2) Calculate the ^{32}P spillover into the ^{3}H window [(2402 – 45)/(80128 – 60)] = 0.0294: This means that 2.94% of all ^{32}P counts are actually being counted in the ^{3}H window, so all ^{32}P counts need to be multiplied by 1.0294. (3) The efficiency of column recovery is now calculated for each sample using the ^{3}H standard (an accounting of the amount of ^{3}H-cAMP that each sample was spiked with before being run through the columns). Each ^{3}H count needs to have both the background (45) and ^{32}P spillover counts subtracted. For sample tube 10 this would be [(41126 – 45) – (1083 × .0294)] = 41049. The column recoveries for the sample data are presented in column III of Table 6.2. (4) We will now calculate the number of counts in the no-membrane control tubes (sample tubes 1 through 3) that will be subtracted from all other corrected counts. Each of the no-membrane sample tubes (1 through 3) can be calculated individually, and the average ^{32}P count for the three tubes will be subtracted from all the other samples in the assay. Sample tube #1 would be calculated as follows:

Step 1, ^{3}H counts: 42719 – 45, ^{32}P counts: 98 – 60.

Step 2, the ^{3}H and ^{32}P counts will now be adjusted for ^{32}P spillover into the ^{3}H window; ^{3}H = 42674 – (38 × 0.0294) = 42673; ^{32}P = 38 × 1.0294 = 39.

Step 3, calculate the column recovery and adjust the ^{32}P count for this recovery; 42673/(52673 – 45) = 0.81 (column recovery), 39/0.81 = 48 corrected ^{32}P counts for tube 1.

If this same calculation is then carried out for the other two no-membrane sample tubes (2 and 3), the average of this triplicate is 48 which is ^{32}P counts subtracted from all other ^{32}P sample counts.

We will now calculate the corrected ^{32}P counts generated for a given sample tube (tube 10): ^{32}P counts of cAMP generated: [(1083 – 60) × 1.0294] ÷ [((41126 – 45) – ((1083 – 60) × 0.0294)) ÷ (52673 – 45)] – 48 = 1302. This corrected ^{32}P count for tube 10 will now be used to calculate the pmol of synthesized cAMP that it represents. The original assay tube contained a trace quantity of ^{32}P-α-ATP and unlabeled ATP in the Rodbell buffer that actually allows the reaction to proceed without limiting substrate. By calculating the specific activity of the ATP in the starting reaction we can calculate the number of moles of ^{32}P-cAMP synthesized. In this sample reaction the final amount of ATP in each tube is 10 nmoles (from the Rodbell buffer). Each tube contained 1,601,360 cpm of ^{32}P-α-ATP (obtained by multiplying the ^{32}P count of IIB (80128 – 60) in Table 6.2 by 20 (since only 1/20 of the standard was counted)). This means that the starting specific activity of ^{32}P-α-ATP in each sample tube was 1,601,360 cpm/10 nmoles. Therefore 1302 cpm (corrected ^{32}P count of sample 10) represents 8.13 pmole of cAMP, and since the tube contained 20 μg protein, our final ^{32}P-cAMP value for tube 10 is 406.5 pmol cAMP/mg membrane protein/time of assay. Figure 6.1 presents the results of the sample assay presented in Tables 6.1 and 6.2. The values in Figure 6.1 represent the average of each triplicate value obtained for each effector.

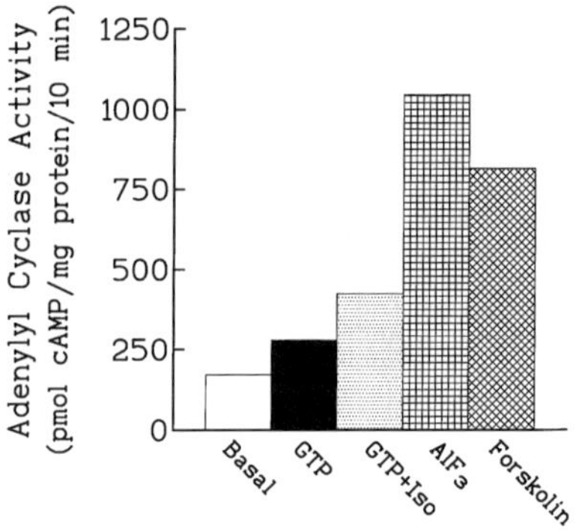

FIGURE 6.1
Representative results of the adenylyl cyclase assay performed in Table 6.1 using the counts generated in Table 6.2. Activities are the average of the triplicate performed for each effector.

F. Limitations of the Adenylyl Cyclase Assay

The measurement of adenylyl cyclase activity in a broken cell preparation offers both advantages and disadvantages. A lysed cell preparation allows access of effectors to each component of the receptor-G protein-adenylyl cyclase cascade. This approach allows an investigator to determine which component of the cascade may be responsible for changes in adenylyl cyclase activity. However, the lysing of a cell releases soluble components, some of which may be critical regulators of adenylyl cyclase function. In studies where loss of soluble regulators may be critical, it may be more desirable to measure cAMP accumulation which is performed in intact cells or tissues using receptor ligands or effectors that can penetrate intact cells (e.g., forskolin).[32]

Many cells contain multiple isoforms of adenylyl cyclase and the amount of synthesized ^{32}P-cAMP measured in this assay is the net effect of activation and perhaps inhibition of multiple isoforms. Insights into the isoform-specific regulation of adenylyl cyclase has required transfection studies (typically in HEK 293 cells) where the transfected subtype overwhelms endogenous adenylyl cyclase activity.[14,28,33-36] Currently, no subtype-specific activators or inhibitors are available which allow selective functional probing of individual adenylyl cyclase isoforms in native cells. Despite these limitations, the ability to measure adenylyl cyclase activity in cellular plasma membrane fractions has greatly expanded the understanding of G protein-coupled signal transduction in general and has allowed insight into the pivotal role that cAMP plays in many cellular processes.

Reagents Needed

4X Rodbell Buffer

200 m*M* Hepes, pH 8.0
200 m*M* NaCl

1.6 m*M* EGTA
40 m*M* $MgCl_2$
0.4 m*M* ATP
0.4 m*M* cAMP
0.25 mg/ml BSA

Stop Buffer (store in 50 ml aliquots at –20°C)

50 m*M* Hepes, pH 7.5
2 m*M* ATP
0.5 m*M* cAMP
2% SDS
1 μl/ml ^{3}H-cAMP (27 Ci/mmol:1 mCi/ml)

2X "hot" Rodbell (prepared the day of assay) recipe for 2 ml (40 tube assay)

1 ml 4X Rodbell
50 μl phosphocreatine (0.56 *M* stock)
50 μl phosphocreatine kinase (35,000 U/ml stock)
8 μl ^{32}P-α-ATP (800 Ci/mmol:10 mCi/ml)
892 μl dH_2O

References

1. Krupinski, J., Coussen, F., Bakalyar, H.A., Tang, W-J., Feinstein, P.G., Orth, K., Slaughter, C., Reed, R.R., and Gilman, A.G., Adenylyl cyclase amino acid sequence: possible channel- or transporter-like structure, *Science,* 244, 1558, 1989.
2. Feinstein, P.G., Schrader, K.A., Bakalyar, H.A., Tang, W.-J., Krupinski, J., Gilman, A.G., and Reed, R.R., Molecular cloning and characterization of a Ca^{2+}/calmodulin-insensitive adenylyl cyclase from rat brain, *Proc. Natl. Acad. Sci. USA,* 88, 10173, 1991.
3. Bakalyar, H.A. and Reed, R.R., Identification of a specialized adenylyl cyclase that may mediate odorant detection, *Science,* 250, 1403, 1990.
4. Gao, B. and Gilman, A.G., Cloning and expression of a widely distributed (type IV) adenylyl cyclase, *Proc. Natl. Acad. Sci. USA,* 88, 10178, 1991.
5. Ishikawa, Y., Katsushika, S., Chen, L., Halnon, N.J., Kawabe, J.-I., and Homcy, C.J., Isolation and characterization of a novel cardiac adenylyl cyclase cDNA, *J. Biol. Chem.,* 267, 13553, 1992.
6. Katsushika, S., Chen, L., Kawabe, J.-I., Nilakantan, R., Halnon, N.J., Homcy, C.J., and Ishikawa, Y., Cloning and characterization of a sixth adenylyl cyclase isoform: Types V and VI constitute a subgroup within the mammalian adenylyl cyclase family, *Proc. Natl. Acad. Sci. USA,* 89, 8774, 1992.
7. Watson, P.A., Krupinski, J., Kempinski, A.M., and Frankenfield, C.D., Molecular cloning and characterization of the type VII isoform of mammalian adenylyl cyclase expressed widely in mouse tissues and in S49 mouse lymphoma cells, *J. Biol. Chem.,* 269, 28893, 1994.

8. Cali, J.J., Zwaagstra, J.C., Mons, N., Cooper, D.M., and Krupinski, J., Type VIII adenylyl cyclase. A Ca^{2+} calmodulin-stimulated enzyme expressed in discrete regions of rat brain, *J. Biol. Chem.,* 269, 12190, 1994.
9. Hellevuo, K., Yoshimura, M., Mons, N., Hoffman, P.L., Cooper, D.M.F., and Tabakoff, B., The characterization of a novel human adenylyl cyclase which is present in brain and other tissues, *J. Biol. Chem.,* 270, 11581, 1995.
10. Paterson, J.M., Smith, S.M., Harmar, A.J., and Antoni, F.A., Control of a novel adenylyl cyclase by calcineurin, *Biochem. Biophys. Res. Comm.,* 214, 1000, 1995.
11. Taussig, R. and Gilman, A.G., Mammalian membrane-bound adenylyl cyclases, *J. Biol. Chem.,* 270, 1, 1995.
12. Tang, W.-J. and Gilman, A.G., Type-specific regulation of adenylyl cyclase by G protein β subunits, *Science,* 254, 1500, 1991.
13. Iwami, G., Kawabe, J., Ebina, T., Cannon, P.J., Homcy, C.J., and Ishikawa, Y., Regulation of adenylyl cyclase by protein kinase A, *J. Biol. Chem.,* 270, 12481, 1995.
14. Kawabe, J., Iwami, G., Ebina, T., Ohno, S., Katada, T., Ueda, Y., Homcy, C.J., and Ishikawa, Y., Differential activation of adenylyl cyclase by protein kinase C isoenzymes, *J. Biol. Chem.,* 269, 16554, 1994.
15. Yoshimura, M. and Cooper, D.M.F., Type-specific stimulation of adenylyl cyclase by protein kinase C, *J. Biol. Chem.,* 268, 4604, 1993.
16. Olianas, M.C. and Onali, P., Participation of delta opioid receptor subtypes in the stimulation of adenylyl cyclase activity in rat olfactory bulb, *J. Pharmacol. Exp. Ther.,* 275, 1560, 1995.
17. Lee, K.H. and McCormic, D.A., Abolition of spindle oscillations by serotonin and norepinephrine in the ferret lateral geniculate and perigeniculate nuclei in vitro, *Neuron,* 17, 309, 1996.
18. Jose, P.A., Raymond, J.R., Bates, M.D., Aperia, A., Felder, R.A., and Carey, R.M., The renal dopamine receptors, *J. Am. Soc. Nephrol.,* 2, 1265, 1992.
19. Childers, S.R. and Deadwyler, S.A., Role of cyclic AMP in the actions of cannabinoid receptors, *Biochem. Pharmacol.,* 52, 819–27, 1996.
20. Patel, Y.C., Greenwood, M.T., Panetta, R., Demchyshyn, L., Niznik, H., and Srikant, C.B., The somatostatin receptor family, *Life Sci.,* 57, 1249, 1995.
21. Shuman, S.L., Capece, M.L., Baghdoyan, H.A., and Lydic, R., Pertussis toxin-sensitive G proteins mediate carbachol-induced REM sleep and respiratory depression, *Am. J. Physiol.,* 269(2 Pt 2), R308, 1995.
22. Kuriyama, K., Hirouchi, M., and Nakayasu, H., Structure and function of cerebral GABAA and GABAB receptors, *Neurosci. Res.,* 17, 91, 1993.
23. Smigel, M.D., Purification of the catalyst of adenylyl cyclase, *J. Biol. Chem.,* 261, 1976, 1996.
24. Strittmatter, S. and Neer, E.J., Properties of the separated catalytic and regulatory units of brain adenylate cyclase, *Proc. Natl. Acad. Sci. USA,* 77, 6344, 1980.
25. Premont, R.T., Chen, J., Ma, H.W., Ponnapalli, M., and Iyengar, R., Two members of a widely expressed subfamily of hormone-stimulated adenylyl cyclases, *Proc. Natl. Acad. Sci. USA,* 89, 9809, 1992.

26. Grammatopoulos, E., Stirrat, G.M., Williams, S.A., and Hillhouse, E.W. The biological activity of the corticotropin-releasing hormone receptor-adenylate cyclase complex in human myometrium is reduced at the end of pregnancy, *J. Clin. Endocrinol. Metab.*, 81, 745, 1996.
27. Hall, I.P., Widdop, S., Townsend, P., and Daykin, K., Control of cyclic AMP levels in primary cultures of human tracheal smooth muscle cells, *Br. J. Pharmacol.*, 107, 422, 1992.
28. Kitten, A.M., Hymer, T.K., and Katz, M.S., Bidirectional modulation of parathyroid hormone-responsive adenylyl cyclase by protein kinase C. *Am. J. Physiol. (Endocrinol. Metab.)*, 266, E897, 1994.
29. Donaldson, J., Brown, A.M., and Hill, S.J., Influence of rolipram on the cyclic 3′,5′–adenosine monophosphate response to histamine and adenosine in slices of guinea-pig cerebral cortex, *Biochem. Pharmacol.*, 37, 715, 1988.
30. Salomon, Y., Londos, C., and Rodbell, M., A highly sensitive adenylate cyclase assay, *Anal. Biochem.*, 58, 541, 1974.
31. Johnson, R.A., Alvarez, R., and Salomon, Y., Determination of adenylyl cyclase catalytic activity using single and double column procedures, *Methods in Enzymology*, 238, 31, 1994.
32. Kelsen, S.G., Higgins, N.C., Zhou, S., Mardini, I.A., and Benovic, J.L., Expression and function of the beta-adrenergic receptor coupled-adenylyl cyclase system on human airway epithelial cells. *Am. J. Respir. Crit. Care Med.*, 152, 1774, 1995.
33. Jacobowitz, O., Chen, J., Premont, R.T., and Iyengar, R., Stimulation of specific types of G_s-stimulated adenylyl cyclases by phorbol ester treatment, *J. Biol. Chem.*, 268, 3829, 1993.
34. Chen, Z., Nield, H.S., Barbier, A., and Patel, T.B., Expression of type V adenylyl cyclase is required for epidermal growth factor-mediated stimulation of cAMP accumulation, *J. Biol. Chem.*, 270, 27525, 1995.
35. Cali, J.J., Parekh, R.S., and Krupinski, J., Splice variants of type VIII adenylyl cyclase, *J. Biol. Chem.*, 271, 1089, 1996.
36. Thomas, J.M. and Hoffman, B.B., Isoform-specific sensitization of adenylyl cyclase activity by prior activation of inhibitory receptors: role of $\beta\gamma$ subunits in transducing enhanced activity of the type VI isoform, *Mol. Pharmacol.*, 49, 907, 1996.

Chapter 7

Methods Used to Assess Specific Messenger RNA Expression During Sleep

Mary Ann Greco, Lalini Ramanathan, Radhika Basheer, and Priyattam J. Shiromani

Contents

0-8493-3361-X/98/$0.00+$.50

The molecular approach to studying biological phenomena has provided key mechanistic insights into many areas of biomedical research. The high specificity and sensitivity of molecular probes make this technology especially attractive. Sleep, however, presents challenges that should be carefully considered both prior to implementing molecular biological techniques and in the conclusions that may be drawn from the results.

One immediate consideration is that it is not clear where in the brain to look for molecular events. Recently, specific mRNAs that are expressed either in sleep or wakefulness have been found. However, these mRNAs were found in the cortex.[1] All of the evidence indicates that sleep and wakefulness are controlled and regulated by hypothalamic–brainstem interactions, the significance of gene expression in the cortex is unclear. How would such expression in the cortex influence the sleep process? Similarly, within the hypothalamus and brainstem, intermingled populations of wake-active and sleep-active neurons are involved in generating sleep-wakefulness. The question is how to separate out these populations and identify expression in wake-active versus sleep-active populations. Procedures such as subtractive hybridization and differential display PCR require homogenization of brain tissue which would obscure the expression in specific neural populations and/or may result in ambiguous data (i.e., a dilution effect), depending on the original sample size. The relative abundance of the marker of interest will, of course, ultimately play a major role in the specific molecular technique chosen. If mRNA expression can be differentially monitored using *in situ* hybridization, it could be the technique of choice for use in sleep studies. On the other hand, if the relative abundance of the mRNA of interest is very low, then semiquantitative RT–PCR is the technique of choice to date. The current *in situ* PCR protocols, while rapidly improving, still require technological refinements before their application to sleep research. In some cases, a combination of molecular techniques (some of which are described in detail below) may yield optimal results.

Another broader consideration is the short, fragmented nature of sleep in rats and mice, animal models of choice in molecular biology studies. In rats and mice, sleep is characterized by short (8 to 10 minutes) bouts of nonREM sleep that are often followed by even shorter (1.5 to 2 minutes) REM sleep periods. This kind of sleep profile can pose problems in experimental designs that seek to elaborate intracellular cascades and thus test specific hypotheses. To investigate the relationship between intracellular events and sleep, we believe it is important to sacrifice animals at specific times (i.e., during nonREM, REM, and waking) without handling the animal. Pompeiano et al.[2] have found increased muscarinic binding during REM sleep. In that study, the rats were killed after individual sleep–wake states without handling them. In our studies, the animals are killed rapidly by means of Nembutal injection into a jugular vein catheter. Using this procedure, we believe that we are able to avoid the effects that handling (i.e., bringing the animal to the guillotine) may have on protein and/or gene expression.

The fragmentary nature of sleep also poses a theoretical question. How would molecular coded events time or regulate the sleep process? For instance, relatively rapid events, such as phosphorylation of proteins (e.g., transcription factors), could be associated with individual sleep–wake bouts. However, it is not clear how events that are relatively slow, such as *de novo* synthesis of protein, might occur during individual sleep or REM sleep bouts.

Another obstacle facing sleep researchers is the lack of an animal model. Areas of neuroscience such as feeding, circadian rhythms, epilepsy, memory, and more recently Alzheimer's disease have benefited from the availability of animal models. Chemical mutagenesis can be used to generate animal models, an approach that has been applied successfully in the area of circadian rhythms.[3] However, in sleep research such an approach is impractical because it would involve screening a large number of animals for signs of aberrant sleep, an intensely labor-intensive process. The approach that we and others are using is that, because we know some of the networking involved in the sleep–wake process, can we identify and utilize existing animal models? This is a practical and mechanistic approach. Tobler et al.[4] recently demonstrated that mice devoid of the prion protein gene have increased arousals and less nonREM sleep (during the second half of the night cycle) compared to wild-type mice. Moreover, the prion protein null mice showed increased slow-wave activity following 6 h total sleep deprivation. These findings have important clinical implications because in the human disease, fatal familial insomnia, a defect in the prion protein gene is suspected to cause the sleep loss.[5] We expect the use of gene knockouts in understanding the sleep process to continue.

The potential caveats described above should be carefully considered when devising protocols that utilize a molecular approach. Molecular approaches are powerful tools and should provide insight into the intracellular events involved in sleep–wake regulation.

In this chapter, the molecular techniques currently used in this laboratory with the rat model system will be described in detail. Northern blot analysis, *in situ* hybridization, and semiquantitative RT–PCR are currently used in this laboratory to monitor steady-state levels of specific mRNAs. In addition to providing information regarding the level of expression of a specific mRNA under experimental conditions, each technique also characterizes the mRNA in a distinctive manner. All the procedures utilize a nucleic acid sequence, or probe, which is chosen to optimize both specificity (its ability to recognize its target) and sensitivity (its detection limit). Since the choice of the type of probe and its visualization can vary greatly over a spectrum of variables (from the target molecule to the experimental system under study), we have included a section devoted to the preparation and use of different types of probes. The technical details described below work very well in our hands, but may need modification when applied to other molecules and model systems. Wherever possible, the reader will be alerted to specific steps which may require further adjustment.

I. Northern Blot Analysis

Northern hybridization is a technique used to determine the amount, size, and integrity of messenger RNA (mRNA), the RNA which encodes protein. In most protocols, total cellular RNA, which contains ribosomal RNA (rRNA), transfer RNA (tRNA), and mRNA, is first size-separated by electrophoresis through agarose that contains formaldehyde. The amount of agarose used will depend on the size of the mRNA of interest (i.e., smaller mRNAs are better separated with a higher percentage agarose). The integrity of the separated RNA is commonly verified by visualization of the larger rRNA subunits (28S and 18S in a 2:1 ratio for "intact" RNA) using acridine orange or ethidium bromide. The RNA is then transferred to a membrane and fixed by vacuum drying (80°C, 2 h) or by UV crosslinkage. The size and relative amount of a specific mRNA species is determined by introduction and subsequent binding of the RNA on the membrane to a specific probe (hybridization). The probe consists of complementary nucleotide sequence to provide specificity which is then marked, or labeled, to permit detection. After hybridization with a radioactive probe, the membrane is exposed to X-ray film. Northern analysis allows detection of a specific mRNA, but it does not provide information on the precise structure of the RNA or the cellular distribution of the message of interest. The protocol below was used to examine steady-state c-fos mRNA levels in different brain regions (Figure 7.1).

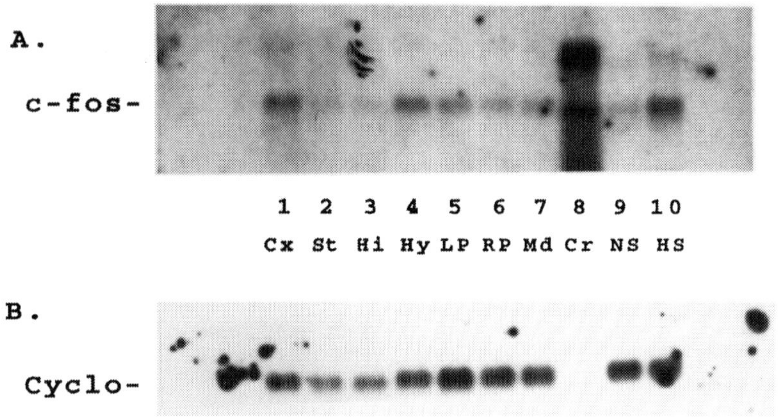

FIGURE 7.1
Steady-state c-fos mRNA levels in rat brain. Total RNA was prepared and analyzed as described in the text. Ten micrograms of total RNA was applied per lane. In A, the membrane is hybridized with the c-fos probe; in B, the membrane is hybridized with the cyclophilin probe. Note that although the same membrane was used for hybridization of both probes, the background present when the riboprobe (cyclophilin) is used is much less than when the double-stranded DNA (c-fos) is used. Note that in lane eight, which contains cerebellum RNA, the c-fos probe hybridizes with two RNA bands and does not hybridize with any size RNA when the cyclophilin probe is used, suggesting that this RNA is partially degraded. The abbreviations are as follows: **Cx,** cortex; **St,** striatum; **Hi,** hippocampus; **Hy,** hypothalamus, **LP,** left pons; **RP,** right pons; **Md,** medulla; **Cr,** cerebellum; **NS,** hypothalamus taken from animal injected with normal saline (0.9%); **HS,** hypothalamus taken from animal injected with hypertonic saline(1.5%).

A. RNA Extraction

RNA is isolated from the samples with TRI reagent. Briefly, this involves homogenizing the tissue in TRI reagent (1ml TRI reagent per 50 to 100 mg tissue), followed by extraction with chloroform, vortexing, centrifuging, and removal of the aqueous phase according to the manufacturer's instructions (Molecular Research Center Inc., Cincinnati, OH). The RNA is precipitated from the aqueous phase with isopropanol (–20°C, at least 2 h, commonly overnight). After vortexing and centrifuging, the RNA pellet is washed with 75% ethanol, air dried, dissolved in sterile water, and stored at –80°C.

B. Gel Electrophoresis

1. Gel composition (1.2% agarose gel): TV = 300 ml

Agarose	3.6 g
Sterile water	246.0 ml
MOPS (10X)[a]	30.0 ml
Formaldehyde (37%)	24.0 ml

[a]10X MOPS, pH 7.0: TV = 4 L

MOPS	167.4 g
EDTA 0.5 *M* stock, pH 8.0	80.0 ml
Sodium acetate, 1 *M* stock	200.0 ml
Sterile water to 4 L	

Note: *The size of the mRNA under examination will determine the percentage agarose used. There is an inverse relationship between the RNA size and the amount of agarose used in electrophoresis. For c-fos mRNA (~2.0 kb, a 1.2% agarose gel was cast).*

2. RNA sample buffer: TV = 775 µl

Formamide (deionized)	500 µl
Formaldehyde (37%)	175 µl
10X MOPS	100 µl

3. 10X loading dye: TV = 16 ml

Glycerol	5 ml
Bromophenol blue	40 mg
Xylene cyanole blue	40 mg
EDTA (10 m*M*)	1 ml
Sterile water	10 ml

Procedure

Rinse the gel apparatus with sterile water. Melt the agarose solution, cool (to ~60°C) before adding MOPS and formaldehyde. Mix well and pour gel under hood. Allow time to polymerize. Prepare RNA samples by adding the appropriate volumes of RNA, 2X volume sample buffer, and 1/10 volume loading dye. Heat the samples at 65°C for 15 min, cool on ice for 5 min, and centrifuge briefly. Load gel carefully. Run gel in MOPS at 100 V for approximately 4 to 5 h. After completion of the run, rinse the gel three times in sterile water to remove formaldehyde. Soak gel in 50 m*M* NaOH for 5 min and again rinse in sterile water. Wet the membrane (Zeta probe GT, Biorad) with sterile water; soak for at least 10 min. The RNA is then transferred in 10X SSC overnight by capillary transfer using a transfer apparatus (Scotlab). The next day, the membrane is placed between 3MM paper (Schleicher and Scheull) and baked in a vacuum oven at 80°C for 1 to 2 h. The membrane is now ready to be hybridized or can be stored dry (away from light) at room temperature.

C. Prehybridization/Hybridization

1. Marker of interest — c-fos

(a) **Type of Probe — double-stranded DNA probe.** The DNA template was obtained by excision of the cDNA insert from the vector by restriction digest and gel purification. The c-fos plasmid, kindly provided by Tom Curran, contains a full-length rat c-fos cDNA (2116 bp) which is inserted into a pSP65 vector.[6] The insert is digested with EcoR1 and purified using Qiaex beads (Qiagen Inc.) after separation by electrophoresis.

(b) **Labeling of Probe — Random priming.** Thaw kit (Pharmacia Biotech) components on ice. While thawing, add sterile water to ~100 ng of DNA template (TV not to exceed 34 µl). Heat at 95°C for 3 min; immediately cool on ice for 3 min. (This will denature the double-stranded DNA.) Then add the components of the kit in the following order to the Eppendorf tube on ice:

Oligolabeling mixture	10 µl
Klenow	1 µl
[^{32}P] dCTP	4 µl

After vortexing and centrifuging briefly, incubate at room temperature for 1 to 2 h. Incubate the membrane in prehybridization solution[b] at the appropriate temperature for 1 to 2 h. Remove the unincorporated nucleotides by passing the mixture through a Nuc-Trap push column (Stratagene) according to the manufacturer's instructions. One µl of post-column solution is counted in a gamma counter to calculate the specific activity of the probe. Heat the post-column eluate at 95°C for 3 min, immediately cool on ice for 3 min, and then add the probe to the prehybridization solution. Hybridization is carried out overnight at the appropriate temperature in a shaking water bath. The next day remove the membrane wash with the following solutions at 65°C:

1. 2X SSC[c]/0.2% SDS	30 min
2. 1X SSC/0.2% SDS	30 min
3. 0.5X SSC/0.2% SDS	2 × 30 min

Expose the membrane to X-ray film (Kodak X-OMAT™) for an appropriate period of time. Expose at –80°C or at room temperature. To strip the probe off the membrane, boil in 0.01X SSC/0.1% SDS for 30 min.

[b]Prehybridization solution (final concentration)

Deionized formamide	50%
Denhardt's (50X stock)	2.5X
SSC (20X stock, see below)	5.0X
SDS (10% stock, wt/vol)	0.5%
Salmon sperm DNA	40 μg/ml

[c]20X SSC, pH 7.0: TV= 1L

Sodium chloride	173.5 g
Sodium citrate	88.2 g

Note: *This protocol works well with the immediate early gene, c-fos. The prehybridization/ hybridization temperature for this probe is 42°C.*

2. Control Marker — cyclophilin (normalizer)

(a) **Type of Probe — RNA probe.** The cyclophilin plasmid, called 1B15, contains a full-length rat cyclophilin cDNA of 700 bp inserted into an pSP65 vector.[7] To generate the antisense sequence required for hybridization, the plasmid is linearized with PstI and transcribed with SP6 RNA polymerase (Promega).

(b) **Labeling of Probe — *In vitro* transcription (Promega).** Thaw kit components on ice. Add components to an Eppendorf tube in the order shown below at room temperature.

Note: *Mixing the components at room temperature prevents precipitation of the template by salts in the buffer which may reduce the transcription efficiency.*

5X transcription buffer	4 μl
100 m*M* DTT	2 μl
Cold UTP	2 μl of 1/1000X diluted
Cold CTP, GTP, ATP	1 μl each
RNasin	1 μl
Linearized DNA template	2 to 3 μl (100 ng)
RNA polymerase	1.5 μl
[^{32}P] UTP	5 μl

After vortexing and centrifuging briefly, incubate the tube for 1 to 2 h at room temperature. Incubate the membrane to be hybridized in the prehybridization solution[d] at the appropriate temperature for 1 to 2 h. After the labeling reaction is complete, remove unincorporated nucleotides as stated above. Add the post-column eluate to the prehybridization solution and incubate at the appropriate temperature overnight in a shaking water bath. The next day remove the membrane from the hybridization solution and wash in the following solutions at 65°C:

1. 0.1X SSC /0.1% SDS	30 min
2. 0.1X SSC/0.1% SDS	60 min
3. 0.05X SSC/0.05% SDS	2 × 2 h

Expose the membrane to X-ray film (X-OMAT, Kodak) for an appropriate period of time at –80°C or at room temperature in the dark. To strip the probe off the membrane, boil the membrane in 0.01X SSC/0.1% SDS for 30 min.

[d]Prehybridization solution (final concentration):

Deionized formamide	50%
SDS	1%
EDTA	2.5 μ*M*
NaCl	1.0 μ*M*
Salmon sperm DNA	10 μg/ml

Note: *This protocol works well with the cyclophilin riboprobe. Prehybridization and hybridization of this probe is done at 65°C. Since riboprobes bind their complementary mRNA species with high affinity, producing intense and sensitive detection with low background, the probes are generally very difficult to strip off and the membrane cannot usually be reused.*

II. *In Situ* Hybridization

In recent years, the *in situ* hybridization (ISH) technique has found widespread application in both basic science and diagnostic clinical research.[8-10] This procedure not only permits quantitation of steady-state mRNA levels, but also provides cellular localization of the message. Detection of mRNA is based on the same principle as described above for Northern hybridization. The probes used for detection can be either single-stranded end-labeled DNA complementary to mRNA (described below) or labeled riboprobe (as described above for cyclophilin probe preparation). The specificity of the probe should be initially tested by Northern hybridization of total RNA, to ensure that the probe recognizes a single mRNA of the correct size. This is necessary because the size specificity of the message of interest cannot be determined by *in situ* hybridization. The following procedure is used in our laboratory for the detection of tyrosine hydroxylase mRNA in the locus coeruleus (LC) of REM-sleep-deprived rats.

A. Fixation

The rats are perfused with 4% paraformaldehyde in phosphate buffer[e] for 6 to 7 min. After removal, brains are immersed in the same fixative for 4 to 6 h. Brain tissue is cryoprotected by rinsing in several changes (~30 min each) of each of the following: 12%, 16%, and 18% sucrose in PBS at 4°C. The brains are finally stored in PBS containing 20% sucrose at 4°C.

[e]PBS, pH 7.4: TV = 1 L

Sodium chloride	9 g
Potassium phosphate (dihydrogen)	0.122 g
Disodium hydrogen phosphate	0.815 g

B. Sectioning

Sections of 14 micron thickness are cut in the cryostat. Sections (3 to 4/slide) are mounted on gelatin/chrome coated slides. Slides are stored at –80°C.

C. Pretreatment of Sections

Prior to hybridization, the stored sections are thawed at room temperature. The slides are treated in the following order:

1. Either 4% paraformaldehyde or 3.7% formaldehyde in PBS	10′
2. PBS (RNase free)	2 × 10′
3. Acetylation[f]	10′
4. PBS (RNase free)	2 × 10′
5. 70% ETOH	1′
6. 90% ETOH	2′
7. 100% ETOH	2′
8. 100% ETOH	5′
9. 90% ETOH	2′
10. Briefly dry the slides	

[f]Acetylation mixture: TV = 200 ml

Triethanolamine	3.71 g
NaCl	1.8 g
10 *N* NaOH	1 ml
Acetic anhydride	500 µl (add just prior to use)

D. *Hybridization/Prehybridization of Slides*

1. **Type of Probe.** A single-stranded oligomeric DNA (30-mer) complementary to TH mRNA (NEN) has been used as probe in this laboratory.

2. **Labeling of Probe.** The oligomer is radiolabeled using a 3′ tailing kit (NEN, NEP-100) with ^{35}S-dATP. The labeling reaction is performed as described by the manufacturer. Mix:

TdT buffer	12.5 μl
$CoCl_2$	2.5 μl
Oligo	10 pmol (TH, 30-mer from NEN)
TdT	2 μl
[^{35}S]dATP	5 μl (50 mCi)

The reaction mixture is incubated at 37°C for 30 min. The probe is purified using a Nensorb 20 column (NEN) as per the manufacturer's instructions; radioactive incorporation is determined using a scintillation counter.

3. **Prehybridization.** Enough prehybridization solution[g] is applied to cover the sections (~200 μl). Leakproof coverslips (Probe Clip PC 200; Grace Bio-Labs) are carefully placed over the sections (avoid trapping any air bubbles). The slides are carefully placed in radioactive boxes. Prehybridization is for 2 hours at 37°C. For hybridization, labeled probe (3×10^6 cpm/slide) is added to the prehybridization solution and mixed well. The coverslips are carefully removed and the prehybridization solution is discarded. Radioactive hybridization solution is added and coverslipped as before. Special care is taken to prevent drying of the sections during this procedure. Hybridization is continued overnight at 37°C.

E. *Washes*

1. Remove the coverslip in 4X SSC containing 10 m*M* β-mercaptoethanol and rinse the slides in the same solution for 5 min at room temperature
2. Wash the slides in 4X SSC for 10 min
3. Wash twice in 2X SSC at room temperature for 10 min each
4. Wash once in 1X SSC at room temperature for 10 min
5. Wash once in 0.5X SSC at 45°C for 20 min
6. Wash once in 0.1X SSC at 45°C for 20 min
7. Slides are air-dried pasted on Whatman paper and exposed to Kodak X-OMAT film. The exposure time depends on the signal intensity. Typically for TH in LC, a 9- to 10-day exposure is used.
8. The sections are subsequently dipped in autoradiographic emulsions, Kodak NTB 2 (for ^{35}S), which is diluted 1:1 with water. The emulsion-coated slides are stored in dark boxes for 3 to 4 weeks at 4°C, developed, and examined under the dark field of a microscope.

[g]***Prehybridization/Hybridization Solution: TV = 25 ml***

50X Denhardt's	2.5 ml
20% SDS	0.25 ml
Formamide	12.5 ml
Water	4.5 ml
20X PIPES[h]	5.25 ml

[h]***20X PIPES, pH 6.8: TV = 500 ml***

NaCl	87.6 g
PIPES	17.12 g
EDTA	18.66 g

The prehybridization mixture can be stored at –20°C. Just prior to use, add salmon sperm DNA (to a final concentration of 250 µg/ml), tRNA (to a final concentration of 250 µg/ml), and DTT (final concentration, 40 m*M*).

Note: *The reagents and protocol described above work well for the detection of tyrosine hydroxylase mRNA in rat brain sections. It is important to note that, while in situ hybridization permits cellular localization of mRNA, only pure antisense strand (DNA or RNA) will provide accurate information. The information obtained can only be verified with the use of the appropriate controls. This eliminates the use of double-stranded DNA as probes in this procedure.*

III. Nonradioactive *In Situ* Hybridization

The use of nonradioactive probes has recently become available. These probes shorten the duration of the procedure (from 2 to 3 weeks to 2 to 3 days), can be stored long term, are safer to handle than radioactive probes, eliminate costly disposal of radioactive waste, and provide optimum resolution. Labeled RNA probes are synthesized by *in vitro* transcription of DNA cloned downstream of SP6, T7, or T3 promoters with the corresponding RNA polymerases, using digoxigenin-labeled UTP as a substrate (instead of ^{32}P- or ^{35}S-UTP). The UTP is linked via a spacer arm to the steroid hapten digoxigenin (DIG-UTP). DIG-labeled RNA or DNA probes are detected after hybridization to target nucleic acid by enzyme-linked immunoassay using an anti-digoxigenin antibody (anti-DIG), followed by a biotinylated secondary antibody and colorimetric detection.

A. *Labeling of Probe*

The following Boehringer Mannheim kit (Genius 4) components are added to an Eppendorf tube on ice:

10X transcription buffer	2 μl
DNA template (linearized)	1 μg
(Add sterile water to a final volume of	13 μl)
10X NTP labeling mixture	2 μl
RNA polymerase	2 μl
RNase inhibitor	1 μl

Vortex, centrifuge briefly, and incubate at 37°C for 2 h. After the labeling reaction is complete, add 1ml of ice-cold absolute ethanol, vortex, and precipitate at –80°C for 1 hour. Centrifuge at 4°C for 20 minutes, wash twice with 75% ethanol, and redissolve in 30 μl of sterile water. The concentration of the probe was determined by comparing the absorbance at 260nm with the labeled control-RNA (vial 5 of the Genius 4 kit). The labeled control RNA contains 10 μg/100 μl of DIG-labeled antisense Neo RNA transcribed with T7 RNA polymerase from 1 μg Pvu II linearized pSPT18-Neo DNA according to the standard protocol. The slides are treated in a manner similar to radioactive *in situ* hybridization. After prehybridization, hybridization, and washing, the slides are subjected to the following (all steps are carried out at room temperature unless otherwise specified):

Day 1

1. Buffer A[i] 2 × 10′
2. Buffer B[j] 1 × 30′
3. Incubate the slides (1:200) anti-DIG (from sheep) in Buffer B overnight at room temperature, coverslipped.

Day 2

1. Wash with Buffer B 2 × 10′
2. Incubate in biotinylated donkey-anti-sheep (1:50) in Buffer B 2 × 60′
3. Wash with Buffer B 2 × 10′
4. Incubate in ABC reagent (Elite PK-6100 standard Vectastain ABC kit) in Buffer B 2 × 60′
5. Wash with Buffer A 2 × 10′
6. Incubate in DAB reagent (Vector SK-4100 DAB kit) variable (depending on amount of mRNA present)
7. Wash with Buffer A 3 × 5′
8. Wash (10′/wash) successively in 30% ethanol, 50% ethanol, 70% ethanol, 90% ethanol, absolute ethanol, xylene. Mount.

[i]Buffer A: 100 m*M* Tris-HCl, pH 7.5/150 m*M* NaCl
[j]Buffer B: 100 m*M* Tris-HCl, pH 7.5/150 m*M* NaCl/3% milk powder

IV. Semiquantitative RT–PCR

Studies designed to assess the expression of mRNAs that are present in low abundance are most sensitively detected by the coupling of reverse transcription (RT) and the polymerase chain reaction (PCR) to one another (RT–PCR). A variety of methods designed to quantitate a specific mRNA by RT–PCR have appeared in the literature since the late 1980s.[12-18] The procedure used in this laboratory utilizes an internal standard which is both reverse transcribed and amplified in the same tube as the marker of interest. Using an internal standard allows one to screen a pool of marker mRNAs from a single total RNA sample in a relatively straightforward manner. The internal standard that has generated the most consistent results in our studies is cyclophilin.[7] Cyclophilin primer sets are designed with the forward and reverse primers of the marker of interest in mind with respect to the potential of primer–dimer formation, the g-c content, and the relative size of the products generated. The procedure described below is best suited to the evaluation of the product ratio (i.e., the marker of interest to the internal standard) across behavioral state. The ultra sensitivity of RT–PCR and the variability of the reagents (i.e., the reverse transcriptase and the Taq polymerase), micropipettes, and hands involved in testing mandates that RT–PCR reactions be performed at least twice with PCR duplicates or triplicates to ensure the reproducibility of the results. The protocol described below was used to investigate the expression of muscarinic receptor subtype mRNAs in the dorsal raphe nucleus across natural sleep (Figure 7.2).

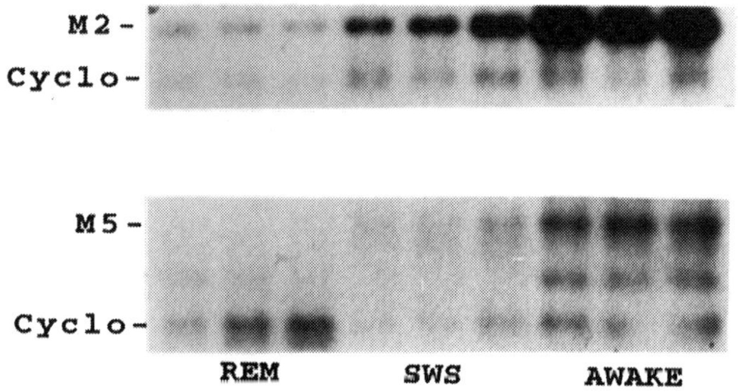

FIGURE 7.2

Steady-state muscarinic receptor levels in the dorsal raphe during natural sleep. The dorsal raphae was removed from rats sacrificed by Nembutal injection into the jugular vein at the times indicated. Purified total RNA was reverse transcribed and amplified as described in the text. The steady-state level of m2 mRNA, whose protein product is a muscarinic receptor, which has been linked with sleep, is lowest during REM sleep and highest during waking. The m5 mRNA exhibits a similar pattern of expression across natural sleep. A physiological role of M5 receptor in sleep has not yet been reported. The PCR product sizes are: 686 bp, m2; 575 bp, cyclophilin; 394 bp, m5; 242 bp, cyclophilin. **Note:** The results presented are in triplicate, one animal per sleep state, and should thus be considered preliminary data.

A. RNA Extraction

Total RNA is extracted from tissue punches as described above (see Northern analysis) with the addition of a DNAse treatment (Gibco/BRL), which is added as a final step. The RNA is quantified spectrophotometrically. One hundred nanograms of total RNA is reverse transcribed per 22 µl RT reaction (see below). If using poly A^+ RNA, 1 to 5 ng is reverse transcribed per 22 µl RT reaction. Note that all RNA preparations are initially tested for DNA contamination by the addition of RNase to the RT mix prior to the initiation of the RT–PCR reaction.

B. Reverse Transcription — TV = 22 µl

1. Precipitation of RNA and Primers (1.5 ml Eppendorf)

Total RNA = 100 ng	
3′ (Reverse) Primers[k] = 50 ng of each	
2 to 3 *M* Sodium acetate, pH (5–6)	10 µl
Glycogen	1 µl (Boehringer-Mannheim)
H_2O	85 µl
ETOH to top	

[k]Cyclophilin reverse primer: 5′–ggtgctctcctgagctacagaagga–3′(based on published cyclophilin cDNA sequence)[7]
M2 reverse primer[19]: 5′–tctgacccgacgacccaacta–3′
M5 reverse primer[19]: 5′–cctgggttgtctttcctgttg–3′
Incubate >20 min at –70°C
Microfuge (at least 20 min), dry pellet
Resuspend in 10 µl H_2O

2. Reverse Transcription

Master Mix (per reaction)
1 µl each dNTP (Stock = 10 m*M*)
2 µl 0.1 *M* DTT (comes with enzyme)
4 µl 5X RT buffer (comes with enzyme)
1 µl Inhibitase (5 Prime-3 Prime)
1 µl SuperScript II Reverse transcriptase (Gibco/BRL)
Vortex briefly; aliquot 12 µl to tube containing RNA
Incubate 90′, 37°C

C. Amplification (TV = 50 µl)

1. Make a Master Mix (per reaction)

1 μl each dNTP (stock 10 m*M*)
5μl 10X PCR buffer (Perkin-Elmer)
50 ng each of forward and reverse primers[1]
H_2O to 45 μl

[1]Cyclophilin oligomers for m2[19]

Forward: 5′ gccgcttgctgcagacatggtcaac 3′
Reverse: see RT reaction

Cyclophilin oligomers for m5[19]

Forward: 5′ caaacacaaatggttcccagt 3′
Reverse: see RT reaction

m2 oligomers[19]

Forward: 5′ cacgaaacctctgacctaccc 3′
Reverse: see RT reaction

m5 oligomers[19]

Forward: 5′ cctgggttgtctttcctgttg 3′
Reverse: see RT reaction

Aliquot 45 μl/tube

Add:

5 μl RT mix
50 μl oil

2. Place the tube in the thermal cycler at 85°C (Hot Start)

Add 0.5μl Amplitaq Polymerase (Perkin-Elmer)

Cycling Conditions

1. 85°C, 4′ (1 cycle)
2. Denature 94°C, 1′
3. Anneal 60°C, 1.5′
4. Extend 72°C, 1.5′

Repeat steps 2 through 4, 25 to 30 cycles

5. Extend 72°C, 10′

Note: *The total number of cycles will vary from marker to marker depending on the relative abundance of each mRNA. Choose a total cycling number which* *ensures that* both markers are synthesized in the linear range. *To check the linearity of two markers, do a pilot experiment in which 5 µl aliquots are removed every 5 cycles after 15 cycles and analyzed by ethidium bromide staining or Southern blot. Extreme care should be taken in oligomer design to pick unique sequences (particularly when working within a gene family) as well as sequences that will not anneal to one another (primer-dimer formation) and will allow stringent annealing temperatures. PCR duplicates or triplicates help eliminate technical errors and make optimal use of the reverse transcription reactions, which should not be stored and reused in any type of quantitative analysis.*

D. *Sample Analysis*

1. Gel electrophoresis: gel composition 3% agarose in TBE buffer[m] (1X), load 5µl PCR reaction/lane.

[m]TBE buffer (20X): TV = 1 L

Tris base	121 g
Boric acid	61.7 g
EDTA	7.44 g

2. Transfer the products from the gel to the membrane (Zetaprobe, Biorad) in 0.4 *M* NaOH. The transfer time is dependent on the product size; transfer of product <500 bp is complete in ~4 h.
3. Prehybridization/Hybridization: TV = 10 ml/membrane

   5X SSC
   10X Denhardt's
   20 m*M* Sodium dihydrogen phosphate, pH 7.0
   7% (wt/vol) SDS
   100 µg/ml salmon sperm DNA

Note: *Prehybridize for at least 20′; hybridization is usually done overnight. The temperature depends on the type of probe used; for double-stranded probes, prehybridization/hybridization reactions are done at 65°C.*

4. Washes

a. 2X SSC/0.1% SDS	2 × 5′, room temperature
b. 0.5X SSC/0.1% SDS	1 × 30′, 65°C
c. 0.1X SSC/0.1% SDS	1 × 30′, 65°C

5. Autoradiographs: use slow film (NEN) without screens if possible. Expose at –80°C.

V. Probes

The choice of a probe and its detection are selected to maximize the signal-to-noise ratio (and therefore provide the greatest specificity and sensitivity) and are important considerations in all the techniques described above (with the exception of double-stranded probes, which should never be used in *in situ* hybridization). The following parameters should be carefully taken into account when choosing a probe:

A. Type of Probe

The three types of probes most frequently used are double-stranded DNA, single-stranded DNA, and single-stranded RNA, all of which may be radiolabeled or nonisotopically labeled.

1. **Double-stranded DNA probes.** These probes, synthesized by nick translation, random-priming, or PCR are commonly used for Northern blot and Southern blot analyses. A probe of this type is optimal to detect mRNAs of moderate to high abundance; other probes may be better for rare mRNAs.[10,11] The major disadvantages of using these probes are 1) the template DNA cannot be removed after labeling, resulting in a reduction of specific activity, and 2) the double-stranded labeled probes will re-anneal over time, decreasing the effective probe concentration.

2. **Single-stranded DNA probes.** These probes may be more useful than the double-stranded probes because they consist of a single antisense strand and can bind target sequence exclusively (and not to the second strand which occurs with double-stranded probes in addition to the target sequence). Single-stranded probes, therefore, do not have to be denatured prior to use in hybridization reactions. These probes may be synthesized by primer extension, linear amplification (asymmetric PCR) or end-labeling of oligonucleotides. In addition, they can be gel purified after synthesis to remove template DNA, providing higher specific activity.

3. **RNA probes.** These probes are synthesized by *in vitro* transcription and have two additional advantages over single-stranded DNA probes. These are: 1) the probe template can be digested by treatment with DNase, avoiding the need to gel purify these probes for Northern analyses, and 2) the higher thermodynamic stability of RNA:RNA duplexes over RNA:DNA duplexes results in stronger hybridization signals with lower background and less cross-hybridization.

B. Probe Specificity

Probes should contain a minimal degree of nonhomologous sequence (i.e., vector and/or intron sequences). Probes should also contain minimal sequence complementarity to closely related genes or highly repeated sequences. Finally, probe sequences should contain as few mismatches as possible.

C. Probe Length

Double-stranded probes that are 100 to 1000 nucleotides in length will generally work well. Increasing the probe length will increase the signal generated; however, probes longer than 1.5 kb may result in a higher background.

D. Types of Labeling

1. Radiolabeling
   a. Double-stranded DNA (random prime labeling): see Northern Analysis section
   b. Single-stranded DNA (end-labeling): see *In situ* section
   c. Single-stranded RNA (*in vitro* transcription): see Northern Analysis section
2. Nonradioactive labeling: see Nonradioactive *In Situ* Hybridization section
3. Specific activity: For radiolabeled probes, the specific activity should be at least 1×10^8 cpm/μg and preferably $>1 \times 10^9$ cpm/μg. Nonisotopically labeled probes should have the maximum degree of substitution that will not interfere with hybridization.

Acknowledgment

This work was supported by funds from the DVA Medical Research and NIH NS30140. Please address all correspondence to Dr. Priyattam Shiromani, VA Medical Center/Harvard Medical School, 940 Belmont Street, Brockton, MA 02401.

References

1. Pompeiano, M., Cirelli, C., and Tononi, G., Changes in gene expression between wakefulness and sleep revealed by mRNA differential display, *Soc. Neurosci. Abstr.*, 22, 688, 1996.
2. Pompeiano, M. and Tononi, G., Changes in pontine muscarinic receptor binding during sleep-waking states in the rat, *Neuroscience Letters*, 109, 347, 1990.
3. Vitaterna, M. H., King, D. P., Chang, A. M., Kornhauser, J. M., Lowley, P. L., McDonald, J.D., Dove, W.F., Pinto, L.H., Turek, S.W., and Takahashi, J.S., Mutagenesis and mapping of a mouse gene, Clock, essential for circadian behavior, *Science*, 264, 719, 1994.
4. Tobler, I., Gaus, S.E., Deboer, T., Achermann, P., Fisher, M., Rulicke, T., Moser, M., Oesch, B., McBride, P.A., and Manson, J.C., Altered circadian activity rhythms and sleep in mice devoid of prion protein, *Nature*, 380, 639, 1996.
5. Gambetti, P., Petersen, R., Monari, L., Tabaton, M., Cortelli, P., and Lugaresi, E., Fatal familial insomnia and the widening spectrum of prion diseases, *Br. Med. Bull.*, 49, 980, 1993.
6. Curran, T., Gordon, M. B., Rubino, K. L., and Sambucetti, L. C., Isolation and characterization of the c-fos(rat) cDNA and analysis of postranslational modification in vivo, *Oncogene*, 2, 79, 1987.

7. Danielson, P. E., Forss-Petter, S., Brow, M. A., Calavetta, L., Douglass, J. , Milner, R. J., and Sutcliffe, G., p1B15: a cDNA clone of the rat mRNA encoding cyclophilin, *DNA*, 7, 261, 1988.
8. Whitfield, H. J. Jr., Brady, L.S., Smith, M.A., Mamalaki, E., Fox, R. J., and Herkenham, M., Optimization of cRNA probe in situ hybridization methodology for localization of glucocorticoid receptor mRNA in rat brain: a detailed protocol, *Cell. Molec. Neurobiol.*, 10, 145, 1990.
9. Valentino, K. L., Eberwine, J. H., and Barchas, J.D., Eds., *In situ Hybridization: Application to Neurobiology,* Oxford University Press, New York, 1987.
10. Chesselet, M-F., Ed., *In Situ Hybridization Histochemistry,* CRC Press, Boca Raton, Florida, 1991.
11. Wilkinson, D.G., Ed., In situ Hybridization. A Practical Approach, IRL Press, New York, 1992.
12. Chelly, J., Kaplan, J.-C., Maire, P., Gautron, S., and Kahn, A., Transcription of the dystrophin gene in human muscle and non-muscle tissues, *Nature (London),* 333, 858, 1988.
13. Wang, A. M. and Mark, D. F., Quantitative PCR, in *PCR Protocols: A Guide to Methods and Applications,* Innis, M. A., Gelfand, D. H., Sninsky, J. J., White, T. J., Eds., Academic Press, 1990, Chap. 9.
14. Gilliland, G., Perrin, S., Blanchard, K., and Bunn, H. F., Analysis of cytokine mRNA and DNA: detection and quantitation by competitive polymerase chain reaction, *Proc. Natl. Acad. Sci. USA*, 87, 2725, 1990.
15. Chelly, J., Montarras, D., Pinset, C., Berwald-Netter, Y., Kaplan, J.-C., and Kahn, A., Quantitative estimation of minor RNAs by cDNA-polymerase chain reaction, application to dystrophin mRNA in cultured myogenic and brain cells, *Eur. J. Biolchem.,* 187, 691, 1990.
16. DiCesare, J., Grossman, B., Katz, E., Picozza, E., Ragusa, R., and Woudenberg, T., A high-sensitivity electrochemiluminescence-based detection system for automated PCR product quantitation, in *Biotechniques,* Perkin-Elmer Corporation, 15, 152, 1993.
17. Vanden Heuvel, J.P., Tyson, F.L., and Bell, D.A., Construction of recombinant RNA templates for use as internal standards in quantitative RT-PCR, in *Biotechniques,* Perkin-Elmer Corporation, 14, 395, 1993.
18. Levesque, G., Lamarche, B., Murthy, M.R.V., Julien, P., Despres, J.-P., and Deshaies, Y., in *Biotechniques,* Perkin-Elmer Cetus Corporation, 17, 738, 1994.
19. Wei, J., Walton, E.A., Milici, A., and Buccafusco, J.J., m1-m5 Muscarinic receptor distribution in rat CNS by RT-PCR and HPLC, *J. Neurochem.,* 63, 815, 1994.

Chapter 8

Competition Binding Assays for Determining the Affinity and Number of Muscarinic Receptor Subtypes in Tissue Homogenates

A. Urban Höglund and Helen A. Baghdoyan

Contents

0-8493-3361-X/98/$0.00+$.50

I. Introduction

Muscarinic receptors play a role in normal physiologic processes such as learning, memory, and arousal.[1-3] Muscarinic receptors are also known to be important in a number of pathophysiologies, including Alzheimer's disease,[4,5] amyotrophic lateral sclerosis,[6] Huntington's disease,[7] and certain psychiatric disorders.[8,9] As a result of their relevance for normal physiology and for disease, muscarinic receptors currently are the focus of considerable research aiming to elucidate the functional consequences of their activation, to specify the mechanisms by which muscarinic receptors themselves are regulated, and to identify and quantify muscarinic receptor subtypes in specific regions of the central nervous system (CNS). This chapter begins with a brief introduction to the molecular and pharmacological classification of muscarinic receptor subtypes. It then presents a detailed protocol useful for identifying and quantifying the muscarinic receptor subtypes present in CNS tissue homogenates. Sample results obtained by using the protocol are provided and interpreted, and limitations of the technique are discussed.

Muscarinic, cholinergic receptors are members of the class of neurotransmitter receptors that mediate responses by interacting with guanine nucleotide binding (G) proteins. The family of muscarinic receptors is comprised of five subtypes which were identified by molecular cloning studies as recently as 1986 (reviewed in references 10 and 11). As a result of their coupling with multiple G proteins, muscarinic receptor subtypes modulate a diverse set of signal transduction pathways (reviewed in references 12 through 14). All five muscarinic receptor subtypes are expressed in mammalian CNS.[15]

Muscarinic receptors are also classified into subtypes based on their interactions with antagonists, as studied using classical radioligand binding techniques (reviewed in references 11 and 16). Pharmacologically identified muscarinic receptors are designated with an upper case M to distinguish them from the molecularly identified (lower case m1–m5) subtypes.[17] In general, the M1–M3 subtypes correspond well to the m1–m3 subtypes.[16,18] The selectivity of muscarinic antagonists for a particular subtype, however, is dose-dependent, and there are no antagonists that have a high affinity for one muscarinic receptor subtype in combination with a low affinity for the other four subtypes.[16] In addition, there are no truly subtype-specific muscarinic agonists.[16,18] As reviewed elsewhere,[1] the lack of subtype-specific muscarinic ligands provides a challenge for *in vivo* studies aiming to determine the roles of individual muscarinic receptor subtypes in mediating behavioral responses.

The pharmacological classification of muscarinic receptor subtypes has been studied in detail.[16,18-20] Briefly, M1 receptors are defined by a high affinity for pirenzepine and a low affinity for methoctramine and himbacine. M2 receptors have a high affinity for methoctramine and himbacine and a low affinity for pirenzepine, 4-DAMP (4-diphenyl acetoxy-methyl piperidine methiodide), and p-F-HHSiD (para-fluoro-hexahydrosiladifenidol). M3 receptors have a high affinity for p-F-HHSiD and 4-DAMP, and a low affinity for pirenzepine.

One important discovery relevant for the pharmacological characterization of muscarinic receptor subtypes is that the antagonist N-methyl scopolamine (NMS) exhibits different dissociation rates from the different muscarinic receptor subtypes.[21] [^{3}H]-NMS dissociates most rapidly from M2 receptors, at an intermediate rate from M1 receptors, and slowly from the M3 and M4 receptors. Most recently, [^{3}H]-NMS dissociation from the m5 subtype has been shown to be the slowest.[22] These differential binding kinetics have been utilized with *in vitro* receptor autoradiography to localize individual muscarinic receptor subtypes throughout primate brain,[23,24] in regions of cat brainstem known to regulate sleep[25] and breathing,[26] and in rat spinal cord,[27] where they may mediate antinociception. The differential dissociation kinetics of [^{3}H]-NMS binding is also useful for studying the allosteric regulation of muscarinic receptor subtypes.[22] By taking advantage of the subtype-selective kinetic differences of [^{3}H]-NMS binding, Waelbroeck and colleagues[28] demonstrated that it is possible to distinguish M1–M4 subtypes in tissue homogenates using competition binding assays. A protocol based on these studies is described below.

II. Protocol

This section presents a detailed protocol for competition of [^{3}H]-NMS binding in tissue homogenates using pirenzepine, AF-DX 116, methoctramine, and methoctramine plus atropine as unlabeled competitors. This protocol is based upon the original studies of Waelbroeck and colleagues.[21,28] These four assays are useful for determining if M1–M4 muscarinic receptor subtypes are present in the selected tissue. The data obtained from these assays can be used to calculate affinity constants and the number of binding sites for M1–M4 receptors. The usefulness of an approach that concurrently assays for multiple muscarinic receptor subtypes is underscored by the knowledge that more than one muscarinic receptor subtype is known to be expressed in many regions of the CNS.[29] The basic pharmacological principles and theoretical considerations underlying these assays are reviewed in the following highly recommended texts (references 30 through 32).

This protocol assumes that 500 mg wet weight of CNS tissue is available so that all four assays can be run concurrently. The protocol can, however, easily be modified to run one assay at a time. The minimum wet tissue weight for one assay is 120 mg. The assays described below are designed for a 30-probe harvester. If a different size harvester is used, then the test tube racks will need to be adjusted accordingly.

A. Buffer, Tissue, and Radioligand Preparation

1. Buffer

Prepare 6 liters of 50 m*M* phosphate buffer containing 1 m*M* $MgCl_2$, pH 7.4. Chill 5.5 liters to 4°C and leave 0.5 liters at room temperature.

2. Tissue preparation
   1. Decapitate a rat and extract the tissue as quickly as possible.
   2. Place the tissue in a petri dish on ice.
   3. Remove the dura and blood vessels.
   4. Determine the wet weight of the tissue. At least 500 mg is needed to run the four competition assays concurrently. Pool tissue from several animals if required.
   5. Homogenize the tissue using 25 ml cold buffer per 500 mg wet weight of tissue.
   6. Centrifuge at 48,000 × *g* for 10 min at 4°C.
   7. Decant the supernatant, resuspend, homogenize, and repeat the centrifugation.
   8. Decant the supernatant, resuspend in distilled water, and take a 0.2 ml sample for protein determination. Centrifuge as before. Commercial kits are available for the protein assay.
   9. Decant the supernatant and add buffer to make a 20 mg/ml dilution. At this point, the tissue homogenate can be frozen and kept at –70°C until assayed. If the assay is planned for the same day, proceed to Step 10.
   10. Measure 40.6 ml buffer into a beaker. Take 24.4 ml of the 20 mg/ml tissue stock and add it to the buffer to make a total of 65 ml. This will be a 7.5 mg/ml tissue concentration, which will produce a final dilution of 3.75 mg/ml.
   11. The tissue homogenate is now ready to be added to the assay tubes. Keep the homogenate on ice until ready to aliquot into assay tubes.
3. Preparation of the radioligand ([^{3}H]-NMS)
   1. To run four assays requires a minimum of 30 ml (120 test tubes × 0.25 ml/tube). Thus, prepare 35 ml of 960 p*M* [^{3}H]-NMS. An example of the calculations to perform for diluting the stock solution of [^{3}H]-NMS follows.
   2. Assume a specific activity for [^{3}H]-NMS of 84 Ci/mmol, supplied as 1 mCi/ml. To calculate the molar concentration of the supplied [^{3}H]-NMS use the formula: (1 mCi/ml)/(84 Ci/mmol) = 11.9 μ*M*.
   3. To dilute the [^{3}H]-NMS stock use the formula: (stock concentration) × (stock volume) = (desired concentration) × (volume). For our example (11.9 mmol/l) × (X ml) = (960 pmol/l) × (35 ml), and X = 2.8 μl. Thus, add 2.8 μl [^{3}H]-NMS stock to 35 ml buffer and vortex.
   4. Take three 100-μl samples of the diluted [^{3}H]-NMS, place in scintillation vials and count. To calculate the expected dpm, use the following formula: expected dpm = $(960 \times 10^{-12}$ mol/l) × (0.1 ml) × $(2.22 \times 10^{12}$ dpm/Ci) × (84 Ci/mmol) = 17,902.
   5. Adjust concentration if necessary.

B. Preparation of Unlabeled Competitor Stock Solutions

1. Atropine sulfate stock solution and serial dilution

To begin, make 2.0 ml of a 400 μ*M* stock solution. The molecular weight of atropine sulfate = 676.8 g/mol. To calculate the amount of atropine needed for the stock solution use the formula: (676.8 g/mol) × (400 × 10^{-6} mol/l) × (2 × 10^{-3} l) = 0.541 mg. Thus, weigh 0.541 mg atropine sulfate and dissolve in 2.0 ml buffer.

To make the serial dilution of the atropine stock solution, begin by labeling two test tubes as follows: 40 μ*M*, 4 μ*M*. Place 9 ml buffer in each tube. Remove 1 ml from the 400 μ*M* stock and add to the test tube labeled 40 μ*M*. Vortex, then remove 1 ml from the 40 μ*M* solution and add to the tube labeled 4 μ*M*.

2. Pirenzepine dihydrochloride stock solution and serial dilution

To begin, make 2.0 ml of a 400 μ*M* stock solution. The molecular weight of pirenzepine dihydrochloride = 424.3 g/mol. To calculate the amount of pirenzepine needed for the stock solution use the formula: (424.3 g/mol) × (400 × 10^{-6} mol/l) × (2 × 10^{-3} l) = 0.339 mg. Thus, weigh 0.339 mg pirenzepine dihydrochloride and dissolve in 2.0 ml buffer.

To make the serial dilution of the pirenzepine stock solution, begin by labeling 11 test tubes as follows: 160 μ*M*, 64 μ*M*, 26 μ*M*, 10 μ*M*, 4.1 μ*M*, 1.6 μ*M*, 655 n*M*, 262 n*M*, 105 n*M*, 42 n*M*, and 17 n*M*. Place 1.5 ml buffer in each of the 11 tubes. Remove 1 ml from the 400 μ*M* stock and add to the test tube labeled 160 μ*M*. Vortex. Repeat this procedure until all dilutions have been made.

3. AF-DX 116 stock solution and serial dilution

AF-DX 116 can be obtained by contacting Boehringer Ingelheim Pharmaceuticals, Inc., 900 Ridgebury Road, P.O. Box 368, Ridgefield, CT 06877-0368, Telephone: (203) 798-9988. To begin, make 2.0 ml of a 400 μ*M* stock solution. The molecular weight of AF-DX 116 = 421.55 g/mol. To calculate the amount of AF-DX 116 needed for the stock solution, use the formula: (421.55 g/mol) × (400 × 10^{-6} mol/l) × (2 × 10^{-3} l) = 0.337 mg. Thus, weigh 0.337 mg AF-DX 116 dihydrochloride and dissolve in 2.0 ml 0.05 *N* HCl.

To make the serial dilution of the AF-DX 116 stock solution, begin by labeling 11 test tubes as follows: 160 μ*M*, 64 μ*M*, 26 μ*M*, 10 μ*M*, 4.1 μ*M*, 1.6 μ*M*, 655 n*M*, 262 n*M*, 105 n*M*, 42 n*M*, and 17 n*M*. Place 1.5 ml buffer in each of the 11 tubes. Remove 1 ml from the 400 μ*M* stock and add to the test tube labeled 160 μ*M*. Vortex. Repeat this procedure until all dilutions have been made.

4. Methoctramine tetrahydrochloride stock solution and serial dilution

To begin, make 2.5 ml of a 400 μ*M* stock solution. The molecular weight of methoctramine = 728.77 g/mol. To calculate the amount of methoctramine needed for the stock solution use the formula: (728.77 g/mol) × (400 × 10^{-6} mol/l) × (2.5 × 10^{-3} l) = 0.729 mg. Thus, weigh 0.729 mg methoctramine tetrahydrochloride and dissolve in 2.5 ml buffer.

To make the serial dilution of the methoctramine stock solution, begin by labeling 11 test tubes as follows: 160 μ*M*, 64 μ*M*, 26 μ*M*, 10 μ*M*, 4.1 μ*M*, 1.6 μ*M*, 655 n*M*, 262 n*M*, 105 n*M*, 42 n*M*, and 17 n*M*. Place 1.5 ml buffer in each of the 11 tubes. Remove 1 ml from the 400 μ*M* stock and add to the test tube labeled 160 μ*M*. Vortex. Repeat this procedure until all dilutions have been made.

C. Competition Binding

The procedure described in this section is identical for each of the three assays to compete [^{3}H]-NMS binding with pirenzepine, methoctramine, and AF-DX 116. The competition assay with methoctramine + atropine differs in that the atropine is added after a 4-h incubation with methoctramine alone. Thus, for the methoctramine + atropine assay, follow all steps below and add only methoctramine as the competitor in Step 3. Use one test tube rack per assay, for a total of four racks. The total assay volume = 1 ml.

1. For each of the four assays, label 30 test tubes (borosilicate glass, size = 13 × 100 mm) as follows: in triplicate, blank; in duplicate, competitor concentrations of 17 n*M*, 42 n*M*, 105 n*M*, 262 n*M*, 655 n*M*, 1.6 μ*M*, 4.1 μ*M*, 10 μ*M*, 26 μ*M*, 64 μ*M*, 160 μ*M*, and 400 μ*M*; in triplicate NSB (nonspecific binding).
2. Dispense 250 μl buffer into the 12 tubes labeled blank.
3. Dispense 250 μl of competitor into the tubes labeled with the corresponding concentration (24 tubes per competitor for a total of 96 tubes).

Note: *For the methoctramine + atropine assay, only methoctramine is added at this time.*

4. Dispense 250 μl of 4 μ*M* atropine into the 12 tubes labeled NSB.
5. Add 250 μl [^{3}H]-NMS (960 p*M*) to all 120 tubes.
6. Add 0.5 ml tissue homogenate (7.5 mg/ml dilution) to all 120 tubes.
7. Incubate at room temperature for 4 h.
8. Extra step for competition with methoctramine plus atropine: After 4 h of incubation add 25 μl of the 40 μ*M* atropine dilution to all 30 tubes in this rack. Vortex and continue the incubation for 35 min. This procedure will displace [^{3}H]-NMS completely from M1 and M2 binding sites, and from 33% of M3 sites and 50% of M4 sites.

D. Terminating the Reaction

1. Preparation of Polyethyleneimidine (0.1%)

Polyethyleneimidine minimizes nonspecific binding of the radioligand to the harvester filters. Place 300 ml buffer in a beaker. Measure 0.6 ml of the 50% (w/v) stock solution and add to the buffer. Stir thoroughly and pour into a flat container suitable for soaking the filters.

2. Harvesting

1. Soak four GF/C filters in 0.1% polyethyleneimidine for at least 2 h.
2. Place 30 test tubes in the harvester rack. These tubes will be used for rinsing the probes.
3. Rinse the harvester vacuum tubes extensively with distilled water.
4. Rinse the vacuum tubes three times with 5 ml of cold buffer.
5. Mount one filter in the harvester and close the apparatus.
6. Rinse the assay test tubes three times with 3 ml of cold buffer. Begin the rinse procedure by filling the test tubes with 3 ml buffer. Rinsing must be performed as quickly as possible to minimize dissociation of [^{3}H]-NMS from the receptors.
7. After the third rinse, remove the filter paper and let dry completely. Using forceps, detach each individual filter and place in scintillation vials.
8. Add 5 ml scintillation solution to each vial. Count the following day.

III. Results and Interpretation

Figure 8.1 shows competition binding curves obtained using rat spinal cord (solid squares) and brain (solid circles). By examining the shape of a competition binding curve, one can determine whether the competitor interacted with the receptor in a simple (i.e., at one binding site) or complex (at more than one binding site) manner. As described in detail by Limbird,[31] a curve of normal steepness proceeds from 90% bound to 10% bound over an 81-fold range of competitor concentration. Normal steepness indicates that the competitor bound to a single species of receptor. In contrast, a shallow competition curve proceeds from 90% to 10% bound over a range of competitor concentration that is greater than 81-fold.[31] Shallow steepness can indicate that the competitor detected multiple, noninteracting binding sites. In Figure 8.1A, the competition curve for spinal cord (solid squares) has normal steepness, indicating that pirenzepine (PZ) detected one population of binding site. Similarly, Figure 8.1B shows that methoctramine (METH) detected one binding site in spinal cord. In contrast, the spinal cord competition curve obtained using methoctramine + atropine (Figure 8.1C; METH + ATR) is shallow, indicating an interaction of methoctramine with more than one receptor population. AF-DX 116 in spinal cord (Figure 8.1D; solid squares) also produced a shallow competition binding curve, demonstrating an interaction with more than one species of binding site.

One goal of the competition binding assays is to identify which muscarinic receptor subtypes are present in the tissue homogenate, and to quantify the relative amounts of these subtypes. Computerized analysis of the competition binding data will provide an estimate of the affinity of the competitor for the receptor (dissociation constant, K_D) and the relative number of binding sites (B_{max}). Detailed reviews

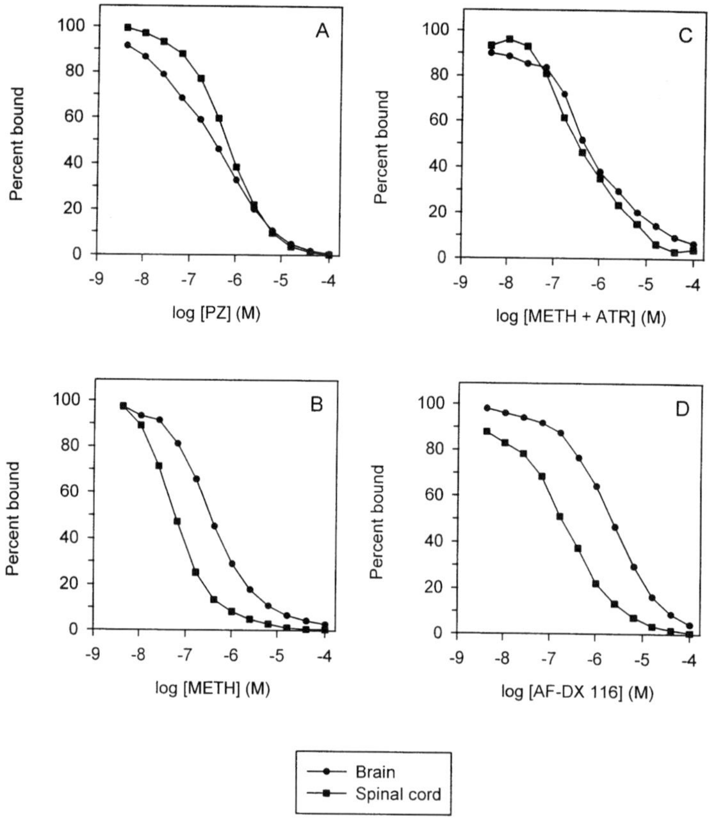

FIGURE 8.1
Competition binding of [^{3}H]-NMS in brain and spinal cord. These graphs plot the percentage of [^{3}H]-NMS bound (ordinate) to muscarinic receptors in rat brain (solid circles) and spinal cord (solid squares) in the presence of increasing competitor concentration (abscissa). Unlabeled competitors shown are pirenzepine (A: PZ); methoctramine (B: METH); methoctramine + atropine (C: METH + ATR); or AF-DX 116 (D). The shape and steepness of these curves provides information about the number of receptor populations present in the tissue, and quantitative analyses of these data yields a measure of the affinity (dissociation constant, K_D) and density of binding sites (B_{max}) for the receptor–competitor interaction.

discussing computerized analysis of binding data are available in references 30 and 33. The K_D values obtained for the antagonists in the tissue homogenate then can be compared to K_D values derived using cloned muscarinic receptor subtypes expressed *in vitro*, or using native receptors expressed in tissues that are enriched in one subtype. This comparison will permit identification of the muscarinic receptor subtypes present in the tissue homogenate. K_Ds for antagonists at known muscarinic receptor subtypes are available from the literature.[11,16,18-20,34]

The sample data shown in Figure 8.1 were analyzed quantitatively using the LIGAND program.[35] A detailed description of muscarinic receptor subtypes identified in rat spinal cord using the protocol presented above recently has been published.[27] Some of the Figure 8.1 results are discussed here to illustrate how these data are interpreted. Competition of [^{3}H]-NMS with pirenzepine in spinal cord (Figure 8.1A; solid squares) revealed that pirenzepine bound one population of receptors with a K_D = 121 n*M*. At human muscarinic receptor subtypes expressed *in vitro*, the n*M* K_Ds of pirenzepine for m1–m4 receptors have been reported to range from 6 to 8 for m1, 224 to 270 for m2, 138 to 150 for m3, and 28 to 37 for m4 receptors.[20,34] Thus, the K_D of pirenzepine derived from the Figure 8.1A data (121 n*M*) indicates an absence of M1 receptors in spinal cord, and the presence of M3 receptors.

Competition of [^{3}H]-NMS with AF-DX 116 in spinal cord (Figure 8.1D; solid squares) showed that AF-DX 116 bound two populations of sites, one site having a relatively high affinity (K_D = 99 n*M*) and one with a lower affinity (K_D = 927 n*M*). At human muscarinic receptor subtypes expressed *in vitro*, the n*M* K_Ds of AF-DX 116 were 1300 for m1, 186 for m2, 838 for m3, and 2800 for m5 receptors.[19] Thus, the Figure 8.1D data demonstrate the presence of M2 (K_D = 99 n*M*) and M3 (K_D = 927 n*M*) receptors in rat spinal cord homogenates.[27]

Figure 8.1 also illustrates how the shape of a competition curve can be altered by the presence of different populations of binding sites. For example, in Figure 8.1A the curve for pirenzepine in brain (solid circles) is shallow compared to the pirenzepine curve in spinal cord (solid squares). The computerized curve fit revealed that in brain, pirenzepine recognized two receptor populations with K_Ds of 12 n*M* and 244 n*M*. These K_Ds correspond to M1 and M2/M3 receptors. Figure 8.1B shows no difference between the methoctramine + atropine curves obtained from brain and spinal cord. This is because competition of [^{3}H]-NMS with methoctramine + atropine recognizes only M3 and M4 receptors, and both of these muscarinic receptor subtypes are known to be present in brain[28] and in spinal cord.[27]

IV. Limitations and Conclusions

One limitation of this protocol is the need for a relatively large amount of tissue. When using tissue from small anatomical structures (i.e., punches of specific brain nuclei), therefore, it is necessary to pool tissue from several animals. Secondly, many of the available computerized data analysis programs provide a less than friendly user interface. Thus, the investigator must invest some time to learn and understand these computer programs.

All studies that use ligand binding techniques to identify and quantify multiple muscarinic receptor subtypes present in the CNS are limited by the lack of truly subtype-specific muscarinic antagonists. Caulfield[16] discusses this issue in detail and makes the point that muscarinic receptor subtypes must be defined using the dissociation constants for a range of subtype selective antagonists. The protocol described in this chapter compensates for this limitation by using more than one subtype selective antagonist, as well as by taking advantage of the differential kinetic properties of [^{3}H]-NMS binding to the different muscarinic receptor subtypes.[21,28] This chapter demonstrates that the K_Ds obtained from this protocol can be interpreted by comparing them with K_Ds for the same antagonists obtained from binding studies performed on cloned receptors expressed *in vitro*, and on native receptors expressed in tissues that are enriched in one subtype (see also reference 27).

Acknowledgments

Supported by the Department of Anesthesia, The Pennsylvania State University College of Medicine, grant B96-99Z-11159-02 from the Swedish Medical Research Council (AUH), and grant MH-45361 (HAB). We thank P. Myers for expert secretarial assistance.

References

1. Baghdoyan, H. A., Cholinergic mechanisms regulating REM sleep, in *Sleep Science: Integrating Basic Research and Clinical Practice,* Monogr. Clin. Neurosci., Schwartz, W. J., Ed., Karger, Basel, 1997, in press.
2. Cuello, A. C., Cholinergic function and dysfunction, *Prog. Brain Res.*, 98, 1, 1993.
3. Lydic, R. and Baghdoyan, H. A., Cholinergic contributions to the control of consciousness, in *Anesthesia: Biologic Foundations*, Biebuyck, J. F., Lynch, C., Maze, M., Saidman, L. J., Yaksh, T. L., and Zapol, W. M., Eds., Raven Press, New York, 1997, in press.
4. Flynn, D. D., Ferrari DiLeo, G., and Mash, D. C., Differential regulation of molecular subtypes of muscarinic receptors in Alzheimer's disease, *J. Neurochem.*, 64, 1888, 1995.
5. McKinney, M. and Coyle, J. T., The potential for muscarinic receptor subtype-specific pharmacotherapy for Alzheimer's disease, *Mayo Clin. Proc.*, 66, 1225, 1991.
6. Berger, M. L., Veitl, M., and Malessa, S., Cholinergic markers in ALS spinal cord, *J. Neurol. Sci.*, 108, 114, 1992.
7. Lange, K. W., Javoy Agid, F., and Agid, Y., Brain muscarinic cholinergic receptors in Huntington's disease, *J. Neurol.*, 239, 103, 1992.
8. Gillin, J. C., Sutton, L., Ruiz, C., Kelsoe, J., Dupont, R. M., Darko, D., Risch, S. C., Golshan, S., and Janowsky, D., The cholinergic rapid eye movement induction test with arecoline in depression, *Arch. Gen. Psychiatry*, 48, 264, 1991.
9. Riemann, D., Hohagen, F., Krieger, S., Gann, H., Müller, W. E., Olbrich, R., Wark, H.-J., Bohus, M., Löw, H., and Berger, M., Cholinergic REM induction test: muscarinic supersensitivity underlies polysomnographic findings in both depression and schizophrenia, *J. Psychiatr. Res.*, 28, 195, 1994.
10. Buckley, N. J., Molecular pharmacology of cloned muscarinic receptors, in *Transmembrane Signalling, Intracellular Messengers and Implications for Drug Development*, Nahorski, S. R., Ed., John Wiley & Sons, New York, 1990, 11.
11. Hulme, E. C., Birdsall, N. J. M., and Buckley, N. J., Muscarinic receptor subtypes, *Annu. Rev. Pharmacol. Toxicol.*, 30, 633, 1990.

12. Baumgold, J., Neurochemical transduction processes associated with neuronal muscarinic receptors, in *CNS Neurotransmitters and Neuromodulators, Acetylcholine*, Stone, T. W., Ed., CRC Press, Boca Raton, 1995, 149.
13. Felder, C. C., Muscarinic acetylcholine receptors: signal transduction through multiple effectors, *FASEB J.*, 9, 619, 1995.
14. Hosey, M., Diversity of structure, signaling and regulation within the family of muscarinic cholinergic receptors, *FASEB J.*, 6, 845, 1992.
15. Levey, A. I., Immunological localization of m1-m5 muscarinic acetylcholine receptors in peripheral tissues and brain, *Life Sci.*, 52, 441, 1993.
16. Caulfield, M. P., Muscarinic receptors — characterization, coupling and function, *Pharmacol. Ther.*, 58, 319, 1993.
17. Birdsall, N., Buckley, N., Doods, H., Fukuda, K., Giachetti, A., Hammer, R., Kilbinger, H., Lambrecht, G., Mutschler, E., Nathanson, N., North, A., and Schwarz, R., Nomenclature for muscarinic receptor subtypes recommended by symposium, *Trends Pharmacol. Sci.*, December Supplement, vii, 1989.
18. Eglen, R. M. and Watson, N., Selective muscarinic receptor agonists and antagonists, *Pharmacol. Toxicol.*, 78, 59, 1996.
19. Buckley, N. J., Bonner, T. I., Buckley, C. M., and Brann, M. R., Antagonist binding properties of five cloned muscarinic receptors expressed in CHO-K1 cells, *Mol. Pharmacol.*, 35, 469, 1989.
20. Dörje, F., Wess, J., Lambrecht, G., Tacke, R., Mutschler, E., and Brann, M. R., Antagonist binding profiles of five cloned human muscarinic receptor subtypes, *J. Pharmacol. Exp. Ther.*, 256, 727, 1990.
21. Waelbroeck, M., Gillard, M., Robberecht, P., and Christophe, J., Kinetic studies of [^{3}H]-N-methylscopolamine binding to muscarinic receptors in the rat central nervous system: evidence for the existence of three classes of binding sites, *Mol. Pharmacol.*, 30, 305, 1986.
22. Ellis, J., Huyler, J., and Brann, M. R., Allosteric regulation of cloned m1-m5 muscarinic receptor subtypes, *Biochem. Pharmacol.*, 42, 1927, 1991.
23. Ferrari-Dielo, Waelbroeck, M., Mash, D. C., and Flynn, D. D., Selective labeling and localization of the M4 (m4) muscarinic receptor subtype, *Mol. Pharmacol.*, 46, 1028, 1994.
24. Flynn, D. D. and Mash, D. C., Distinct kinetic binding properties of N-[^{3}H]-methylscopolamine afford differential labeling and localization of M1, M2, and M3 muscarinic receptor subtypes in primate brain, *Synapse*, 14, 283, 1993.
25. Baghdoyan, H. A., Mallios, V. J., Duckrow, R. B., and Mash, D. C., Localization of muscarinic receptor subtypes in brain stem areas regulating sleep, *NeuroReport*, 5, 1631, 1994.
26. Mallios, V. J., Lydic, R., and Baghdoyan, H. A., Muscarinic receptor subtypes are differentially distributed across brain stem respiratory nuclei, *Am. J. Physiol.*, 268, L941, 1995.
27. Höglund, A. U. and Baghdoyan, H. A., M2, M3, and M4, but not M1, muscarinic receptor subtypes are present in rat spinal cord, *J. Pharmacol. Exp. Ther.*, 281, 470, 1997.

28. Waelbroeck, M., Tastenoy, M., Camus, J., and Christophe, J., Binding of selective antagonists to four muscarinic receptors (M1 to M4) in rat forebrain, *Mol. Pharmacol.*, 38, 267, 1990.
29. Vilaro, M. T., Mengod, T. G., and Palacios, J. M., Advances and limitations of the molecular neuroanatomy of cholinergic receptors: the example of multiple muscarinic receptors, *Prog. Brain Res.*, 98, 95, 1993.
30. Hulme, E. C., Ed., *Receptor-Ligand Interactions*, Oxford University Press, New York, 1992, 458.
31. Limbird, L. E., *Cell Surface Receptors: A Short Course on Theory and Methods*, Second Ed., Martinus Nijhoff Publishing, Boston, 1996, 238.
32. Yamamura, H. I., Enna, S. J., and Kuhar, M. J., Eds., Methods in *Neurotransmitter Receptor Analysis*, Raven Press, New York, 1990, 267.
33. Unnerstall, J. R., Computer-assisted analysis of binding data, in *Methods in Neurotransmitter Analysis*, Yamamura, H. I., Enna, S. J., and Kuhar, M. J., Eds., Raven Press, New York, 1990, 37.
34. Bolden, C., Cusack, B., and Richelson, E., Antagonism by antimuscarinic and neuroleptic compounds at the five cloned human muscarinic cholinergic receptors expressed in Chinese hamster ovary cells, *J. Pharmacol. Exp. Ther.*, 260, 576, 1992.
35. Munson, P. J. and Rodbard, D., LIGAND: a versatile computerized approach for characterization of ligand-binding systems, *Anal. Biochem.*, 107, 220, 1980.§

Chapter 9

Isolation and Identification of Specific Transcripts by Subtractive Hybridization

Thomas S. Kilduff, Luis de Lecea, Hiroshi Usui, and J. Gregor Sutcliffe

Contents

0-8493-3361-X/98/$0.00+$.50

I. Applications of Subtractive Hybridization in Neuroscience

The phenotypic differences between cell types are determined by the genes they express. Consequently, a number of molecular methods have been developed to reveal differential gene expression within and between tissues in various stages of development or physiological conditions. Since its introduction[1] and early success in the identification of the T-cell receptor and other immune-related genes,[2-4] the subtractive hybridization method has been a very productive technique for the isolation of mRNAs specific to tissue types and stages of development. Of particular interest to neuroscientists have been the applications to identify transcripts specific to the cerebral cortex,[5-8] striatum,[9] hypothalamus,[10] visual system,[11,12] olfactory system,[13] auditory system,[14] peripheral CNS,[15] glial cells,[16,17] development of the CNS,[18-21] and in pathological CNS conditions.[22-25] This method has also been used to identify genes involved in the circadian system of *Neurospora*[26] and, along with expression cloning,[27] was successfully used to isolate the long-sought mRNA encoding serotonin N-acetyl-transferase from the pineal gland,[28] the rate-limiting enzyme for melatonin biosynthesis. Subtraction hybridization has also been applied to isolate molecules associated with sleep deprivation,[29] and at least one of these molecules has been characterized further.[30] More recently, a subtractive approach led to the identification of cortistatin, a neuropeptide with a high degree of homology to somatostatin that is specific to cortex and hippocampus and which appears to modulate slow-wave activity measured by the cortical electroencephalogram.[31]

II. Subtractive Hybridization: Important Considerations

A. Theoretical Issues

Subtractive hybridization provides a means of depleting nucleotide sequences which are common to two mRNA populations, resulting in a relative enrichment of sequences unique to the experimental target tissue of interest. Hybridization of cDNA made from polyA+ RNA isolated from the target tissue occurs in the presence of excess polyA+ mRNA or cRNA made from the control tissue which acts as the "driver" in the hybridization reaction. Whereas the bulk of nucleotide sequences will hybridize to their complements, sequences that are unique between the two populations will remain single-stranded and can be separated from the double-stranded RNA populations by column chromatography.

The success of any subtractive hybridization approach will be determined in large part by two characteristics of the mRNA population in the biological system to be investigated. First, the extent to which the *abundance* of a specific transcript differs between tissue types or within a tissue between experimental conditions. A second important parameter is the *nucleotide sequence complexity* of the mRNA populations in the tissues to be investigated. Both parameters are classically identified from the reassociation kinetics of mRNA to single-copy DNA. Based on reassociation kinetics, mRNAs can be classified into high (>1000 molecules per cell), medium (20 to 1000 molecules per cell), and low (<20 molecules per cell) abundance classes. In eukaryotic cells, the majority of the 0.5 to 1.5×10^5 transcripts fall into the low abundance class and tissue-specific transcripts may be expected to be of very low abundance. The CNS is a particularly challenging tissue because of its biochemical complexity and large percentage of low abundance transcripts. Of the approximately 30,000 brain mRNAs, as many as 20,000 are estimated to be brain-specific, but the number of alternatively spliced mRNAs ensures that the actual number of distinct mRNAs must exceed these values.[32]

Among methods for screening cDNA libraries, differential screening with classical cDNA probes labeled to high specific activity appears to be suitable primarily for the detection of high abundance mRNAs, down to approximately 0.1 to 0.01% of the mRNA population. A higher level of sensitivity, estimated to be at least 0.005%, is afforded by screening standard cDNA libraries with subtracted cDNA probes, in which probes are enriched for specific transcripts through hybridization to an mRNA population lacking the sequences of interest. The same enrichment procedure used to generate a subtracted probe can also be used to create subtracted libraries which can increase the representation of low abundance mRNAs (<0.001%) and thereby greatly decrease the number of clones which must be screened to encounter a rare transcript. However, subtracted libraries are often of poor quality due to degradation of single-stranded DNA on hydroxyapatite during the separation step.

B. mRNA Sources for Subtractions

In the early applications of this technique, polyA+ RNA from the experimental tissue of interest (target) was reverse-transcribed into single-stranded cDNA and hybridized with 10- to 100-fold excess of polyA+ RNA from the control (driver) tissue in a small volume (5 to 10 µl). This classic procedure would yield unhybridized cDNA suitable both for use as a subtracted cDNA probe and as a source for subsequent cloning into a subtracted library, if desired. However, the procedure was very costly in terms of polyA+ RNA for the hybridization step, because it might require 30 to 200 µg polyA+ RNA for use as driver in the hybridization reaction. As a point of comparison, about 20 to 40 µg polyA+ RNA is typically obtained from a whole adult rat brain, clearly precluding the application of the classical approach to small brain regions. In contrast, the directional tag PCR subtraction procedure described in this chapter and illustrated in Figure 9.1 calls first for creation of cDNA libraries from the tissues of interest. Once good-quality cDNA libraries exist, a virtually unlimited amount of cRNA can be made for use as a driver in the hybridization step.

Another issue for consideration is the mRNA source for the construction of cDNA libraries. Perhaps the majority of investigators who make cDNA libraries start with total cellular RNA from which poly A+ selection is performed. Poly A+ selection from such a source includes mature cytoplasmic, membrane-bound, and free polysomal RNA and thus would be likely to have a considerable amount of

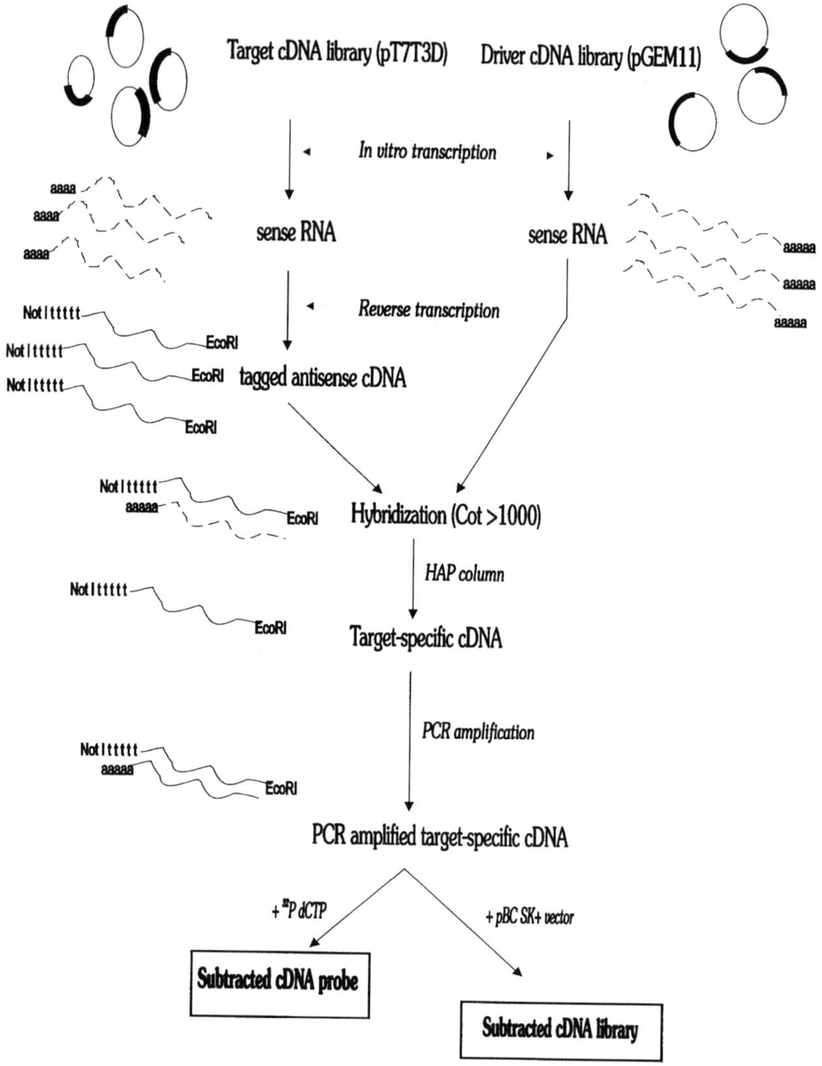

FIGURE 9.1
Schematic of the subtractive hybridization procedure employed by Gautvik et al.[10] and described in the text.

incompletely processed polyadenylated mRNAs. Although this may not be an important consideration for most applications, it can be particularly troublesome in subtractive hybridization, which is essentially an enrichment procedure. In theory, all mRNAs that differ between the experimental and control tissues will be enriched by the subtraction and, consequently, any transcript, whether completely processed or not, may be disproportionately represented in the subtracted library. Consequently, it has been our experience that isolation of cytoplasmic RNA prior to poly A+

selection yields superior results because fewer false positives are detected during subsequent screening steps. Unfortunately, this procedure precludes the possibility of using RNA isolated from frozen tissue as a source.

C. Example of a Successful Subtractive Hybridization: Isolation of Hypothalamus-Specific mRNAs

Although there are now numerous examples of successful applications of subtractive hybridization to the CNS as indicated above, the protocol given below is slightly modified from that used in a recent study in which the hypothalamus was the target of the subtraction. The existence of a large number of brain-specific genes indicates that a substantial portion of the mammalian genome is dedicated to the function of the central nervous system.[32] Among those genes that have been identified, there are now multiple examples of genes which can be considered as hypothalamus-specific, vasopressin and oxytocin being the prototypical examples. The hypothesis tested by Gautvik et al.[10] was that mRNAs are selectively expressed in discrete hypothalamic nuclei to encode proteins which subserve functions unique to this brain region. As mentioned above, key technical aspects of the subtractive hybridization which led to its success are: 1) isolation of cytoplasmic RNA from fresh tissue as a source for poly A+ selection, and 2) construction of directional cDNA libraries as sources of cDNA and cRNA for the target and driver, respectively, in the subtraction. Equally important, though not explicitly mentioned above, are: 1) choice of the driver tissue(s) in the hybridization reaction, and 2) a logical series of screens to eliminate potential false positives. In this particular application, two rounds of subtraction were conducted with driver RNAs isolated from different tissues: hippocampus and cerebellum. Although both of these tissues are of neural origin, the biochemistry of these brain regions are distinct from each other and from the hypothalamus. The procedures used in the hypothalamus subtraction and the protocol given below are derived in large part from previous work by Usui et al.[9] The general approach is illustrated in Figure 9.1.

III. A Detailed Protocol for Subtractive Hybridization

A. *RNA Isolation and cDNA Library Construction*

Young adult rats of both sexes were decapitated and the brains dissected into hypothalamus, cerebellum, and hippocampus. Cytoplasmic RNA was rapidly isolated and poly A selected using oligo(dT)-cellulose. The hypothalamic target library was constructed from 2 μg of poly A+ mRNA using the pT7T3D vector (Pharmacia; Piscataway, NJ). A NotI- oligo dT adaptor-primer (from the Directional Cloning Toolbox, Pharmacia; Piscataway, NJ) was used for first-strand synthesis and an EcoRI adaptor (also from the Directional Toolbox) was ligated into the double-stranded cDNA . The hippocampus and cerebellum driver libraries were constructed from 2 μg of poly A+ in pGEM11Zf(–) (Promega; Madison, WI). The subtracted library was constructed in pBCSK+ (Stratagene; La Jolla, CA) because lower backgrounds were found than when pT7T3D was used.

B. Preparation of Target cDNA

In this portion of the protocol, the cDNA library made from the target tissue is linearized by restriction endonuclease digestion, transcribed into cRNA, and then the cRNA serves as a template for synthesis of radiolabeled cDNA.

1. Digest the target cDNA library with Not I:

pT7T3D library DNA (at 1.0 μg/μl)	20 μl (20 μg)
10X H buffer (Boehringer Mannheim; Indianapolis, IN)	30 μl
H_2O	248 μl
Not I (10 U/μl)	2 μl

Incubate at 37°C for 2 h. Check digestion by examining 1 μl in an agarose gel.

2. Extract with phenol-$CHCl_3$.
3. Precipitate the DNA by addition of 1/2 vol of 7.5 *M* NH_4OAc and 2.5 vol of ethanol (30 min at –20°C).
4. Dissolve the DNA in 20 μl of DEPC-treated water. Determine OD_{260} of a 1 μl aliquot.
5. *In vitro* transcription reaction

H_2O (DEPC)	12 μl
10X Transcription buffer (Megascript; Ambion; Austin, TX)	4 μl
75 m*M* rATP	4 μl
75 m*M* rCTP	4 μl
75 m*M* rGTP	4 μl
75 m*M* rUTP	4 μl
Not I digested library	4 μl (2 μg)
T7 RNA polymerase mix	4 μl

Incubate at 37°C for 3 h.

6. Add 1.5 μl RNase-free DNase (Ambion; Austin, TX). Incubate at 37°C for 15 min.
7. Extract with phenol-$CHCl_3$ once. Add 1/2 vol of 7.5 *M* NH_4OAc to the aqueous phase and precipitate the RNA with 3 vols of EtOH. Spin 5 min.
8. Resuspend the pellet in 90 μl of DEPC H_2O and add 10 μl of 10X buffer H (or any other restriction enzyme buffer). Add 1 μl of RNase-free DNase I. Incubate for 20 min at 37°C.

9. Extract with phenol-$CHCl_3$ once. Add 1/2 vol of 7.5 *M* NH_4OAc to the aqueous phase and precipitate the RNA with 3 vols of EtOH. Spin again to remove supernatant. Resuspend in 100 µl of DEPC H_2O.

10. cDNA synthesis (Antisense cDNA with tag sequence)

Dilute 2 µg of cRNA in 10 µl of DEPC H_2O. Heat 10 min at 68°C, then chill on ice. Spin and add 10 µl of α-[^{32}P]-dCTP.

Mix in a new tube:

First strand mix (Pharmacia; Piscataway, NJ)	11 µl
DTT	1 µl
Oligo dT Not primer (dilute 5µg/µl stock to 1:4)	1 µl (=1 µg)
Heat denatured cRNA	20 µl

Incubate at 37°C for 1 h.

11. Alkaline treatment.

a. Add:

0.5 *M* EDTA (pH 8.0)	1.6 µl
10% SDS	1.6 µl

Mix well

b. Add 5 µl of 3 *N* NaOH and incubate the mixture at 68°C for 30 min.

c. Cool the mixture to room temperature. Add 12 µl of 1 *M* Tris.Cl (pH 7.4). Mix well and add 6 µl of 2 *N* HCl.

12. Extract with phenol-$CHCl_3$ once. Back-extract with 50 µl of TE. Pool both fractions together.

13. Separate unincorporated label by Sephadex G-50 spun column chromatography.

C. *Preparation of the Driver cRNA*

Digestion of the driver cDNA library and *in vitro* transcription proceeds in a manner similar to that described above for the target cDNA library. Note, however, the larger volume of the restriction digest and the use of 3 *M* NaOAc in Step 3 instead of 7.5 *M* NH_4OAc.

1. Digest the driver cDNA library with Not I:

pGEM11 library DNA (at 1.0 µg/µl)	20 µl (20 µg)
10X H buffer (Boehringer Mannheim; Indianapolis, IN)	40 µl

H_2O	256 µl
Not I (10 U/µl)	5 µl

Incubate at 37°C for 2 h. Check for complete digestion by running 1 µl in an agarose gel.

2. Extract with phenol-$CHCl_3$ twice and then once with $CHCl_3$.
3. Precipitate the DNA by addition of 1/10 vol of 3 *M* NaOAc and 2.5 vol of ethanol.
4. Dissolve the DNA in 50 µl of DEPC-treated water. Determine OD_{260} of a 1 µl aliquot.
5. *In vitro* transcription reaction

H_2O (DEPC)	11 µl
10X Transcription buffer (Megascript, Ambion; Austin, TX)	4 µl
75 m*M* rATP	4 µl
75 m*M* rCTP	4 µl
75 m*M* rGTP	4 µl
75 m*M* rUTP	4 µl
Not I digested library	5 µl (3 µg)
T7 RNA polymerase mix	4 µl

Incubate at 37°C for 3 h.

6. Add 2 µl RNase-free DNase (Ambion; Austin, TX). Incubate at 37°C for 15 min.
7. Extract with phenol-$CHCl_3$ once and precipitate with 1/2 vol of NH_4OAc and 3 vols of EtOH. Resuspend the pellet in 90 µl of H_2O and add 10 µl of 10X buffer H (or any other restriction enzyme buffer).
8. Add 1 µl of RNase-free DNase I and incubate at 37°C for 15 min. Extract with phenol-$CHCl_3$ once and precipitate with 1/2 vol of NH_4OAc and 3 vols of EtOH. Resuspend the pellet in 100 µl of H_2O.
9. Determine OD using a 1 µl aliquot.

D. Subtractive Hybridization and Hydroxyapatite (HAP) Column Separation

In this portion of the protocol, the target cDNA is hybridized to excess driver cRNA for 24 h at 68°C and then the single-stranded nucleotide sequences are separated from the double-stranded hybrids by hydroxyapatite chromatography. Details on how to run the jacketed column in Steps 11 and 12 are given in standard molecular biology manuals such as in Appendix E of Sambrook et al.[33]

1. Phenol/chloroform extract the cDNA target. Keep 1/10 (~10 µl) of the cDNA for use as a control cDNA probe and for alkaline gel electrophoresis.

2. Mix the remaining 9/10 volumes of cDNA target with 20 to 50 μg of the driver RNA and precipitate the mixture by adding 1/10 vol 3 *M* NaOAc and 3 volumes of ethanol.
3. Dissolve the pellet in 5 μl DEPC-H_2O. Add:

Solution A (see *Reagents Needed*)	4 μl
DEPC-treated 5 *M* NaCl	1 μl

4. Transfer the mixture to a 1.5 ml screw cap Sarstedt tube and overlay with 50 μl of mineral oil.
5. Boil the mixture for 90 seconds and incubate at 68°C for 24 h.
6. Add 100 μl of chloroform and 90 μl of TE-0.5 *M* NaCl. Mix well and spin.
7. Extract the aqueous phase with phenol-$CHCl_3$ and precipitate the target DNA and driver RNA by adding 2.5 vol of ethanol.

Note: *No salt used here because of the high concentration used in the previous step.*

8. Dissolve the pellet in 0.5 ml of 50 m*M* phosphate buffer (see *Reagents Needed*) (pH 6.5)/0.2%SDS.
9. Add 50 m*M* PB/0.2% SDS to the hydroxyapatite (HAP) powder (HTP Hydroxylapatite resin; BioRad; Emeryville, CA). Boil for 30 min. (Preheating reduces nonspecific binding).
10. Allow the suspension to settle at room temperature and aspirate the buffer over the settled material. Decant again after resuspension of the HAP in 5 to 10 ml of 50 m*M* PB/0.2% SDS. Resuspend settled resin in 5 ml of 50 m*M* PB/0.2% SDS.
11. Fill the column with 4 to 5 ml of PB 50 m*M*/0.2%SDS. Apply the slurry with a pipette into a water-jacketed column (Bio-Rad; Emeryville, CA) connected to a 60°C circulating water bath. Connect the peristaltic pump.
12. Wash the column with 5 ml of preheated (60°C) 50 m*M* PB/0.2% SDS at a flow rate of 0.5 ml/min.
13. While prewashing, mix the pellet of target and driver by pipetting up and down. Add 0.5 ml of the sample preheated to 60°C. As soon as the sample is loaded, begin to collect 0.5 ml fractions in Eppendorf tubes. (About 40 drops for a fraction collector).
14. After loading, wash the column with 2 mls of preheated 50 m*M* PB/0.2%SDS three times. Continue collecting 0.5 ml fractions.
15. Wash the column with 2 mls of preheated 120 m*M* PB/0.2% SDS. (3×, total 6 ml)
16. Wash the column with 2 mls of preheated 400 mM PB/0.2% SDS. (3×, total 6 ml)
17. Count the radioactivity of each tube in Cerenkov mode (with the whole tube). Figure 9.2 provides typical expected results.
18. Collect 4 tubes from the 120 m*M* phosphate concentration.
19. Extract with phenol-$CHCl_3$ in 15 ml Falcon tubes. Save the supernatant and concentrate and desalt each sample with Centricon 100 (Amicon; Beverly, MA) spun at 1000 × *g* (= 2620 rpm in H600 Sorvall rotor) at room temperature.
20. Measure the volume after Centricon (normally 35 μl) and add H_2O to 100 μl.

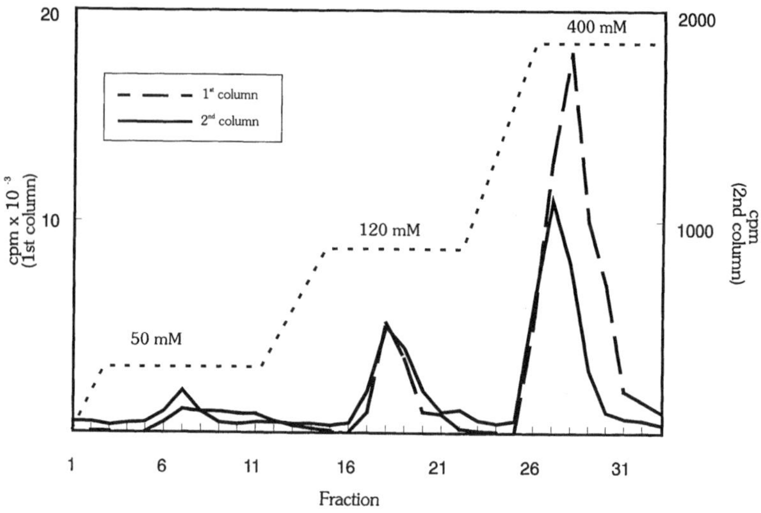

FIGURE 9.2
Idealized elution profiles obtained after subtractive hybridization. Low molecular weight fragments elute after 50 m*M* phosphate buffer is added to the column; single-stranded enriched cDNA elutes at 120 m*M*; and the remaining double-stranded material containing the majority of the radioactivity elutes at 400 m*M*. Two rounds of subtraction are illustrated here; note the difference in the labels on the ordinate axes.

21. Add:

0.5 *M* EDTA pH 8.0	4 µl
20% SDS	2 µl
2 *N* NaOH	18 µl

Incubate for 30 min at 68°C.

22. Cool mixture at room temperature and add:

Tris.Cl pH 7.5	40 µl
2 *N* HCl	12 µl
5 *M* NaCl	7 µl (= 1/25 vol)
Glycogen or driver cRNA (see *Reagents Needed*)	5 µl
EtOH	500 µl

Note: *If you are doing a second round of subtraction use 30 to 50 µg of cRNA and go back to Step 3. If it is already the second round, use glycogen as a carrier.*

23. Precipitate at –80°C for 30 min. Spin, discard supernatant, and dry pellet. Resuspend in 10 µl H_2O. Use a 1/10 dilution for PCR.

E. Synthesis of the Subtracted Probe

Having isolated the single-stranded cDNA population in the previous step, a subtracted probe is synthesized for differential screening of the target, driver, and subtracted cDNA libraries.

1. PCR amplification

Template 1/10 dilution of the 120 m*M* fraction (or diluted control)	1 µl (1 µl)
10X A-PCR buffer (see *Reagents Needed*)	10 µl
DMSO	5 µl
dNTP labeling mix (2.5 m*M* d(A,G,T)TP, 0.25 m*M* dCTP)	4 µl
Primer set (300 to 600 ng each HUT7T3 and HUNOPRIM*)	2 µl
H_2O	67 µl
α–[^{32}P] dCTP	10 µl

Heat at 94°C for 5 min.

Add 0.5 µl of Taq polymerase (Perkin Elmer; Foster City, CA).

PCR for 30 cycles at 94°C for 15 sec; 60°C for 15 sec; 72°C for 1 min.

2. Extract with phenol-$CHCl_3$ once.
3. Purify probe by Sephadex G-50 spun column chromatography or Microspin S-200.
4. More than 10^6 cpm/ml should be used for hybridization.

* HUT7T3D: 5′–AACTGGAAGAATTCGCGG–3′
HUNOPRIM: 5′–AGGCCAAGAATTCGGCACGA–3′

F. Construction of the Subtracted cDNA Library

If the subtraction procedure has been successful, nucleotide sequences common to the target and driver mRNA populations should be depleted and sequences unique to the target should be enriched in the subtracted cDNA library. The following protocol calls for pooling of the products of three ligation reactions before transformation.

1. PCR amplify without isotope. Extract with phenol-$CHCl_3$. Prepare a SizeSep 400 Sephacryl S400 column equilibrated in ligation buffer (same procedure as used in library construction using Time Saver kit; Pharmacia; Piscataway, NJ).
2. Load the PCR reaction on the column. Spin at 400 *g* for 2 min.

3. Add 2 µl Not I and incubatc for 1 h. Add 1 µl EcoRI and incubate another 1 h at 37°C.
4. Heat inactivate at 68°C and phenol-$CHCl_3$ extract.
5. Load on SizeSep 400 Spun column (Pharmacia; Piscataway, NJ) pre-equilibrated in ligation buffer.
6. Set up three different ligation reactions:

Column eluate	25 µl	15 µl	10 µl
1X ligase buffer*	—	10 µl	15 µl
PEG buffer*	6 µl	6 µl	6 µl
Vector pT7T3D (50 ng/µl)	2 µl	2 µl	2 µl
diluted ATP (1/20)*	1 µl	1 µl	1 µl
T_4 DNA ligase	1 µl	1 µl	1 µl

Incubate at 16°C for 1 to 2 h.

7. Pool the three ligations, extract with phenol-$CHCl_3$, and precipitate with NH_4OAc and EtOH. Resuspend pellet in 5 µl of H_2O.
8. Transform 1 µl from the ligation into high efficiency electrocompetent SURE (Stratagene; La Jolla, CA) cells.

IV. Closing Comments

This protocol describes the procedures for conducting the subtraction, generating a subtracted probe, and creating a subtracted library. Of course, the success of any subtraction in leading to the identification of clones of interest will in large part depend upon a carefully thought-out series of screens to limit the number of potential candidates to pursue further. The hypothalamus subtraction study[10] employed 1) hybridization to Southern blots made from cDNAs contained within the target and driver libraries, 2) partial DNA sequencing of selected cDNAs, and 3) assessment of tissue distribution by Northern blots and *in situ* hybridization in the screening process subsequent to the subtraction. Although these methods may also be used fruitfully to screen candidates in many other applications of subtraction hybridization, the most appropriate procedures to use in the screening process will vary with the goal of the specific experiment.

There are a number of other variations of the subtractive procedure described here. One popular variation calls for labeling the driver RNA with photobiotin and hybridization of the target and driver RNAs in the presence of streptavidin.[34] A recent variation involves binding the driver poly(A)+ RNA pool to paramagnetic Dynabeads Oligo (dT)25[35] to allow removal of the sequences in common after hybridization.

* From Time-Saver kit (Pharmacia; Piscataway, NJ).

Other investigators have attempted to incorporate a control for all steps in the cDNA enrichment procedure, thereby enabling measurement of specific vs. nonspecific hybridization and subtraction in the same tube.[36] Suppression PCR has also been integrated into subtractive hybridization resulting in a new variation, termed *suppression subtractive hybridization* (SSH), which combines normalization and subtraction in a single procedure. In this procedure, the normalization step equalizes the abundance of cDNAs within the target population and the subtraction step excludes the common sequences between the target and driver populations.[37] An examination of the annual number of references in Medline in which *subtractive hybridization* is used as a keyword attests to the utility of this procedure and ensures that the method will continue to evolve.

Reagents Needed

Solution A:

125 m*M* HEPES pH (7.6)	0.5 *M* stock	50 µl
5 m*M* EDTA DEPC-treated	0.5 *M* stock	2 µl
0.5% SDS	10% stock	10 µl
H_2O		138 µl

Phosphate buffer:

Prepare a 1 *M* stock of Na_2HPO_4 and another 1 *M* stock of NaH_2PO_4. Combine equal volumes of both stocks to make a 1 *M* Na Phosphate stock solution. This stock is normally slightly acidic but will have the appropriate pH when diluted.

10X APCR buffer:

67 m*M*	Tris.Cl pH 8.8
167 m*M*	Ammonium sulfate
67 m*M*	$MgCl_2$
0.1%	Gelatin

Acknowledgments

We gratefully acknowledge the contributions of Monica Carson, Patria Danielson, Pam Foye, Kaare Gautvik, and Elizabeth Thomas in the development of the protocol to its current state. Research described herein has been supported in part by the National Institutes of Health (AG11084, GM32355 and NS33396) and the Army Research Office (DAAH04-95-1-0616).

References

1. Zimmermann, C. R., Orr, W. C., Leclerc, R. F., Barnard, E. C., and Timberlake, W. E., Molecular cloning and selection of genes regulated in Aspergillus development, *Cell*, 21, 709, 1980.
2. Hedrick, S. M., Cohen, D. I., Nielsen, E. A., and Davis, M. M., Isolation of cDNA clones encoding T cell-specific membrane-associated proteins, *Nature*, 308, 149, 1984.
3. Chien, Y., Becker, D. M., Lindsten, T., Okamura, M., Cohen, D. I., and Davis, M. M., A third type of murine T-cell receptor gene, *Nature*, 312, 31, 1984.
4. Davis, M. M., Cohen, D. I., Nielsen, E. A., Steinmetz, M., Paul, W. E., and Hood L., Cell-type-specific cDNA probes and the murine I region: the localization and orientation of Ad alpha, *Proc. Natl. Acad. Sci. USA*, 81, 2194, 1984.
5. Travis, G. H. and Sutcliffe, J. G., Phenol emulsion-enhanced DNA-driven subtractive cDNA cloning: isolation of low-abundance monkey cortex-specific mRNAs, *Proc. Natl. Acad. Sci. USA*, 85, 1696, 1988.
6. Travis, G. H., Naus, C. G., Morrison, J. H., Bloom, F. E., and Sutcliffe, J. G., Subtractive cloning of complementary DNAs and analysis of messenger RNAs with regional heterogeneous distributions in primate cortex, *Neuropharmacology*, 26, 845, 1987.
7. Watson, J. B., Battenberg, E. F., Wong, K. K., Bloom, F. E., and Sutcliffe, J. G., Subtractive cDNA cloning of RC3, a rodent cortex-enriched mRNA encoding a novel 78 residue protein, *J. Neurosci. Res.*, 26, 397, 1990.
8. Bernal, J., Godbout, M., Hasel, K. W., Travis, G. H., and Sutcliffe, J. G., Patterns of cerebral cortex mRNA expression, *J. Neurosci. Res.*, 27, 153, 1990.
9. Usui, H., Falk, J. D., Dopazo, A., de Lecea, L., Erlander, M. G., and Sutcliffe, J. G., Isolation of clones of rat striatum-specific mRNAs by directional tag PCR subtraction, *J. Neurosci.*, 14, 4915, 1994.
10. Gautvik, K. M., de Lecea, L., Gautvik, V. T., Danielson, P. E., Tranque, P., Dopazo, A., Bloom, F. E., and Sutcliffe, J. G., Overview of the most prevalent hypothalamus-specific mRNAs, as identified by directional tag PCR subtraction, *Proc. Natl. Acad. Sci. USA*, 93, 8733, 1996.
11. Smith, D. P., Shieh B. H., and Zuker, C. S., Isolation and structure of an arrestin gene from Drosophila, *Proc. Natl. Acad. Sci. USA*, 87, 1003, 1990.
12. Higashide, T., Murakami, A., McLaren, M. J., and Inana, G., Cloning of the cDNA for a novel photoreceptor protein, *J. Biol. Chem.*, 271, 1797, 1996.
13. Dear, T. N., Boehm T., Keverne E. B. and Rabbitts T. H., Novel genes for potential ligand-binding proteins in subregions of the olfactory mucosa, *Embo J*, 10, 2813, 1991.
14. Robertson, N. G., Khetarpal, U., Gutierrez-Espeleta, G. A., Bieber, F. R., and Morton, C. C., Isolation of novel and known genes from a human fetal cochlear cDNA library using subtractive hybridization and differential screening, *Genomics*, 23, 42, 1994.
15. Akopian, A. N. and Wood, J. N., Peripheral nervous system-specific genes identified by subtractive cDNA cloning, *J. Biol. Chem.*, 270, 21264, 1995.
16. Rhyner, T. A., Lecain, E., Mallet, J., and Pessac, B., Isolation of cDNAs from a mouse astroglial cell line by a subtracted cDNA library, *J. Neurosci. Res.*, 27, 144, 1990.

17. Baba, H., Fuss, B., Watson, J. B., Zane, L. T., and Macklin, W. B., Identification of novel mRNAs expressed in oligodendrocytes, *Neurochem. Res.*, 19, 1091, 1994.
18. Miller, F. D., Naus, C. C., Higgins, G. A., Bloom, F. E., and Milner, R. J., Developmentally regulated rat brain mRNAs: molecular and anatomical characterization, *J. Neurosci.*, 7, 2433, 1987.
19. Watson, J. B. and Margulies, J. E., Differential cDNA screening strategies to identify novel stage-specific proteins in the developing mammalian brain, *Dev. Neurosci.*, 15, 77, 1993.
20. Prasad, S. S. and Cynader, M. S., Identification of cDNA clones expressed selectively during the critical period for visual cortex development by subtractive hybridization, *Brain Res.*, 639, 73, 1994.
21. Wijnholds, J., Chowdhury, K., Wehr, R., and Gruss, P., Segment-specific expression of the neuronatin gene during early hindbrain development, *Dev. Biol.*, 171, 73, 1995.
22. Abe, K., Sato, S., Kawagoe, J., Lee, T. H., and Kogure, K., Isolation and expression of an ischaemia-induced gene from gerbil cerebral cortex by subtractive hybridization, *Neurol. Res.*, 15, 23, 1993.
23. Duguid, J. R., Rohwer, R. G., and Seed, B., Isolation of cDNAs of scrapie-modulated RNAs by subtractive hybridization of a cDNA library, *Proc. Natl. Acad. Sci. USA*, 85, 5738, 1988.
24. Krady, J. K., Oyler, G. A., Balaban, C. D., and Billingsley, M. L., Use of avidin-biotin subtractive hybridization to characterize mRNA common to neurons destroyed by the selective neurotoxicant trimethyltin, *Brain. Res. Mol. Brain. Res.*, 7, 287, 1990.
25. Munoz, A., Rodriguez-Pena, A., Perez-Castillo, A., Ferreiro, B., Sutcliffe, J. G., and Bernal, J., Effects of neonatal hypothyroidism on rat brain gene expression, *Mol. Endocrinol.*, 5, 273, 1991.
26. Loros, J. J., Denome, S. A., and Dunlap, J. C., Molecular cloning of genes under control of the circadian clock in Neurospora, *Science*, 243, 385, 1989.
27. Coon, S. L., Roseboom, P. H., Baler, R., Weller, J. L., Namboodiri, M. A., Koonin, E. V., and Klein, D. C., Pineal serotonin N-acetyltransferase: expression cloning and molecular analysis, *Science*, 270, 1681, 1995.
28. Borjigin, J., Wang, M. M., and Snyder, S. H., Diurnal variation in mRNA encoding serotonin N-acetyltransferase in pineal gland, *Nature*, 378, 783, 1995.
29. Rhyner, T. A., Borbely, A. A., and Mallet, J., Molecular cloning of forebrain mRNAs which are modulated by sleep deprivation, *Eur. J. Neurosci.*, 2, 1063, 1990.
30. Neuner-Jehle, M., Rhyner, T. A., and Borbely, A. A., Sleep deprivation differentially alters the mRNA and protein levels of neurogranin in rat brain, *Brain Res.*, 685, 143, 1995.
31. de Lecea, L., Criado, J. R., Prospero-Garcia,, O., Gautvik K. M., Schweitzer, P., Danielson, P. E., Dunlop, C. L., Siggins, G. R., Henriksen, S. J., and Sutcliffe, J. G., A cortical neuropeptide with neuronal depressant and sleep-modulating properties, *Nature*, 381, 242, 1996.
32. Sutcliffe, J. G., mRNA in the mammalian central nervous system, *Annu. Rev. Neurosci.*, 11, 157, 1988.
33. Sambrook, J., Fritsch, E. F., and Maniatis, T., *Molecular Cloning A Laboratory Manual*, 2nd ed., Cold Spring Harbor Laboratory Press, Cold Spring Harbor, NY, 1989.

34. Welcher, A. A., Torres, A. R., and Ward, D. C., Selective enrichment of specific DNA, cDNA and RNA sequences using biotinylated probes, avidin and copper-chelate agarose, *Nucleic Acids Res.*, 14, 10027, 1986.
35. Meszaros, M. and Morton, D. B., Subtractive hybridization strategy using paramagnetic oligo(dT) beads and PCR, *Biotechniques*, 20, 413, 1996.
36. Gastel, J. A. and Sutter, T. R., A control system for cDNA enrichment reactions, *Biotechniques*, 20, 870, 1996.
37. Diatchenko, L., Lau, Y. F., Campbell, A. P., Chenchik, A., Moqadam, F., Huang, B., Lukyanov, S., Lukyanov, K., Gurskaya, N., Sverdlov, E. D., and Siebert, P. D., Suppression subtractive hybridization: a method for generating differentially regulated or tissue-specific cDNA probes and libraries, *Proc. Natl. Acad. Sci. USA*, 93, 6025, 1996.

Chapter 10

Use of *In Situ* Hybridization Histochemistry to Study Muscarinic Receptor mRNA Expression in Brains of Sleep-Deprived Rats

Clete A. Kushida and Donna M. Simmons

Contents

0-8493-3361-X/98/$0.00+$.50

I. Introduction

In situ hybridization histochemistry (ISHH) is a flexible molecular biologic technique which can be used to identify the distribution of receptors in organs and tissues. ISHH results in an association between any two polynucleotide chains, provided a relationship of base complementarity exists between these chains. The expression of messenger RNA (mRNA) is localized histologically in tissue sections by application of a radioactively labeled probe (e.g., ^{35}S-UTP) containing sequences which are complementary to the mRNA to be localized.[1]

ISHH localizes to the cell body and identifies cells synthesizing the receptors, which is an advantage over conventional radioligand autoradiography and immunohistochemistry. The latter two techniques typically provide morphologic data about labeled cells by identifying the final location of receptor molecules. Since these molecules are frequently transported to dendrites or axons, they can be found distant from the cell bodies.[1a] Additionally, newly synthesized muscarinic receptors in cell bodies and axons may also be en route to their terminal destinations.[2] Other advantages of ISHH include: 1) the ability to obtain quantitative assessments of receptor mRNAs as well as qualitative mapping of their distribution,[3] 2) less reliance on membrane properties (e.g., fluidity) which may affect ligand binding in experimental conditions,[4] and 3) facilitation of the study of regulatory changes following experimental manipulation.[5]

We used the technique of *in situ* hybridization histochemistry to study REM sleep deprivation effects on m1–m3 muscarinic receptor mRNA expression in the rat brain.[6] Since cholinergic agonists and antagonists alter REM sleep,[7] muscarinic receptors are implicated in REM sleep regulation. Muscarinic receptors are classified by their affinity for different cholinergic antagonists. Four pharmacologically distinct receptor subtypes (M1–M4)[8,9] and five different receptor genes or cDNAs (m1–m5) have been identified.[10] Binding studies of these cloned receptors expressed into various cell lines show that m1, m2, and m3 correspond to the pharmacologic M1, M2, and M3 receptors, respectively;[11] m4 probably represents the M4 receptor;[12] and no pharmacologic equivalent for m5 has been identified. These receptors were localized in the rat brain by radioligand autoradiography, immunohistochemistry, and, most recently, by *in situ* hybridization histochemistry (ISHH). Since the technique of ISHH offers greater specificity compared to ligand binding methods, we used this comparatively newer technique to study the effects of REM-sleep deprivation on the expression of m1, m2, and m3 muscarinic receptor mRNA in the rat brain.

II. Protocol

The following protocol uses procedures of basic molecular biologic research protocols and has been extensively tested with the collaboration of Drs. Rebecca Zoltoski and J. Christian Gillin. Plasmid constructs containing cDNA coding for the specific I3 regions of the muscarinic receptors were obtained from T. I. Bonner.[13]

A. Glassware and Solution Preparation

All glassware should be washed with Alconox glassware detergent, thoroughly rinsed with distilled water, and then soaked in DPC-treated water overnight. DPC-treated water is prepared by adding 10 µl of 0.001% diethyl pyrocarbonate per liter of distilled water. The glassware should be autoclaved. Any solution, if not sterile, should also be autoclaved. The quantity of the solutions described below are computed for one standard glass slide dish containing one slide rack holding 30 microscope slides. Approximately 300 ml of solution will cover the slides in the slide rack. In addition, an adult rat's brain from the cervical spinal cord–medullary junction to the anterior tip of the forebrain will occupy approximately 300 slides, each containing five 20 µm-thick sections.

B. Animal Perfusion

Standard pH-shift perfusion with a 4% paraformaldehyde solution is used for brain immunohistochemistry.[14]

C. Slide Preparation

The frozen-sectioned brain tissue is placed on slides coated with a solution of gelatin, chrome alum, and poly L-lysine which promote tissue adherence and also decrease nonspecific probe binding.

1. **Preparation of Subbed Slides** (gelatin, distilled water, chromium potassium sulfate). New microscope slides are placed in slide racks using semisterile techniques. The slides are soaked for a minimum of 30 min in hot tap water with strong Alconox glassware detergent. The slides should be thoroughly rinsed in hot tap water and then distilled water for 30 min each. The slide racks should be briefly drained and then immersed into the subbing solution for 2 min. The slide racks should be briefly removed from the subbing solution, and then reimmersed in the same solution for 3 min. The subbing solution is made with 5 g gelatin and 500 ml distilled water. Heat and stir to dissolve, but do not heat above 50°C. Add 0.5 g chromium potassium sulfate (chrome alum), filter, and use while warm. When immersing the slides in the subbing solution, avoid creating bubbles, since they will leave ridges or outlines in the gelatin on the slides. The slide racks should be drained of subbing solution in a dust-free area and stored for at least 2 h at 37 to 50°C. The slides should then be allowed to cool before storage in dust-free slide boxes.
2. **Lysine Coating** (poly L-lysine, 10 m*M* Tris pH 8.0). The subbed slides should be loaded in slide racks using semisterile techniques. Place the subbed slides in a solution of 25 mg poly L-lysine and 500 ml of 10 m*M* Tris pH 8.0 for 20 to 30 min. Drain the slide racks and store at 37°C for at least 30 min. The slides should then be allowed to cool before storage in dust-free slide boxes. The slides are usable for 4 to 6 weeks.

D. Preparation of Tissue Sections

Following removal, blocking, and freezing, the brain is sectioned in a cryostat and the sections are placed directly on the prepared slides. The brains are covered with a standard embedding medium and allowed to freeze at –15 °C in the cryostat. Frozen sections of 20 μm thickness are obtained from the perfused rat brains. Approximately five sections are mounted onto the subbed, poly L-lysine coated slides. Gloves should be worn at all times to avoid ribonuclease contamination. A 22-mm-wide coverslip will subsequently cover each slide, so the slides should each have a tissue-free border of 2 mm. The sections should be allowed to air dry and then be stored at least 2 h in a vacuum with desiccant. The sections should then be stored at –20°C and used within 24 h, otherwise there will be an increase in nonspecific background hybridization. The following steps should be conducted using sterile technique, sterile reagents, and all glassware should be rinsed with DPC-treated water to inhibit ribonuclease.

E. Prehybridization Treatment

With fixed tissue, the mRNA may be unavailable to the probe; thus, tissue deproteination by proteinase K increases probe access. The tissue is placed in a buffer solution, acetylated to block the positive charges from the lysine, rinsed, and then dehydrated with ethanol. Each of the 300-ml solutions described below is designed to fill a standard slide dish containing a full rack of microscope slides.

1. **Proteinase K** (1 *M* Tris pH 8.0, 50 m*M* EDTA pH 8.0, 10 μg/ml proteinase K, distilled water). The slides are immersed in a 0.001% proteinase K solution for 30 min at 37°C. A 300-ml solution can be prepared, using 30 ml 1 *M* Tris pH 8.0, 30 ml 0.5 *M* EDTA pH 8.0, 300 μl 10 mg/ml proteinase K (1% in distilled water), and 240 ml distilled water.
2. **TEA Buffer** (0.1 *M* triethanolamine pH 8.0, distilled water). The slides are rinsed in 0.1 *M* triethanolamine (TEA) at pH 8.0 for 2 to 3 min. A 300-ml solution of this buffer is prepared from 60 ml 0.5 TEA pH 8.0 and 240 ml distilled water.
3. **Acetylation** (acetic anhydride, 0.1 *M* TEA pH 8.0). The slides are acetylated for 10 min at room temperature. To a dry slide dish, 0.75 ml acetic anhydride and 300 ml 0.1 *M* TEA pH 8.0 are added prior to adding the slides.
4. **SSC Rinse** (sodium chloride, sodium citrate). The slides are placed in a slide dish containing 2X SSC, prepared from a stock solution of 20X SSC (175.4 g sodium chloride and 88.2 g sodium citrate per liter of distilled water). The slides should be immersed in this solution three times for 2 min each, with fresh SSC each time.
5. **Dehydration** (50%, 70%, 95%, and 100% ethanol). The slides are dehydrated in ascending ethanol concentrations: 50% ethanol for 3 min, 70% ethanol for 3 min, 95% ethanol for 3 min, 100% ethanol for 3 min, and new 100% ethanol for 3 min. The slides are drained and stored with desiccant under vacuum at room temperature for at least 2 h until hybridization. The slides should not be stored for longer than 24 h, since nonspecific background hybridization will increase.

F. *Riboprobe Preparation*

Using the RNase-free, linearized DNA as a template, a complementary radioactively labeled RNA transcript is obtained. The specific activity of the probe is obtained by a scintillation counter. The probe is purified by removing the DNA template with RNase-free DNase I. The probe is then extracted with phenol and chloroform; ethanol precipitation and size-exclusion chromatography are used to remove the majority of unincorporated nucleotides and to concentrate the RNA.

1. **DNA Template Linearization** (DPC-treated water, reaction buffer, plasmid DNA, EcoRI, Tris chloride/EDTA buffer, phenol chloroform, chloroform, 3 *M* sodium acetate pH 5.2, 70% and 100% ethanol) . For the study of m1 receptor mRNA, combine 12 μl DPC-treated water, 2 μl reaction buffer, 4 μl of M1 plasmid DNA (400 μg/μl), and 1 μl EcoRI. For the study of m2 receptor mRNA, combine 8 μl DPC-treated water, 2 μl reaction buffer, 8 μl of M2 plasmid DNA (600 μg/μl), and 2 μl EcoRI. For the study of m3 receptor mRNA, combine 14 μl DPC-treated water, 2 μl reaction buffer, 2 μl of M3 plasmid DNA (1.025 μg/μl), and 2 μl EcoRI. The total EcoRI/DNA solution should be 20 μl in volume. After heating for 30 min at 37°C, 2 μl of EcoRI should be added to the solution, and the solution should be heated for another 30 min. Approximately 130 μl of a Tris chloride/EDTA buffer should be added to the solution. An equal volume of 150 μl of phenol chloroform should be added, and the mixture is vortexed for 30 sec and then spun in a centrifuge at 5000 rpm for 2 min. The EcoRI precipitate should then be discarded, and another 150 μl of phenol chloroform should be added to the top layer. This mixture is vortexed for 30 sec, spun in a centrifuge at 5000 rpm for 2 min, and the EcoRI precipitate is discarded. One hundred microliters of chloroform is added to the mixture; it is vortexed for 30 sec and spun for 2 min. The top layer is retained and approximately 3 *M* sodium acetate at pH 5.2 is added to the solution, with a total volume of 165 μl. Three volumes, or 495 μl (165 μl × 3) of ice-cold 100% ethanol is added. The mixture is vortexed and placed in –70°C for 15 min. The mixture is then centrifuged for 8 min; the pellet is retained and washed with 300 μl 70% ethanol. The pellet is suspended in solution by gentle agitation, and then vortexed for 30 sec. Following centrifugation for 10 min at 5000 rpm, the pellet is retained and heated in a 37°C water bath for 15 min.
2. **Riboprobe Transcription** (RNase inhibitor, 5X transcription buffer, 100 m*M* DTT, 10 m*M* rCTP, 10 m*M* rATP, 10 m*M* rGTP, ^{35}S-UTP, SP6 polymerase, DNase I, Na/TrisCl/EDTA). The pellet is resuspended by adding 1.5 μl of RNase inhibitor (40,000 units/ml), 4 μl of 5X transcription buffer, 0.5 μl 100 m*M* DTT, 1 μl 10 m*M* rCTP, 1 μl 10 m*M* rATP, and 1 μl 10 m*M* rGTP. The mixture is gently agitated and then 10 μl ^{35}S-UTP (specific activity = 1000 Ci/mmol) is added. Lastly, 1 μl SP6 polymerase is added to the mixture. The mixture is centrifuged at 5000 rpm and then placed in a 37°C water bath for 30 min. One microliter of SP6 polymerase is added to reboost the enzymatic activity. After an additional 30 min in the water bath, the DNA is eliminated by adding 1.0 μl DNase I and 1.5 μl RNase inhibitor. The mixture is placed in the 37°C water bath for 10 min. Size-exclusion chromatography is conducted by adding approximately 70 μl of Na/TrisCl/EDTA to a column filter containing nonorganic fibers. Approximately 46.5 μl of Na/TrisCl/EDTA is added to the mixture to bring the volume up to 70 μl, and this mixture is added to the column. An additional

70 μl of Na/TrisCl/EDTA is added to the column. The solution at the bottom of the column should be collected, and a 1 μl sample is removed for scintillation counting.

3. **Scintillation Counts.** At least 5×10^6 cpm/ml is obtained. The following formulas are used to calculate the volume of the riboprobe needed for the study:

$$H = 0.075 \text{ ml} \times S$$
$$V = (5 \times 10^6 \text{ cpm/ml}) \times H \times (1 \text{ μl/R})$$

where H = total volume of hybridization solution, S = number of slides, V = total volume of riboprobe needed for the study, and R = cpm from scintillation counter.

G. Hybridization

The RNA probe solution is mixed with hybridization buffer, applied to the tissue and allowed to incubate overnight. Approximately 75 μl of the hybridization solution is required for each microscope slide, using a 22 × 60 mm coverslip.

1. **Preparation of Hybridization Solution A** (formamide, 5 *M* NaCl, 1 *M* Tris pH 8.0, 0.5 *M* EDTA pH 8.0, 100X Denhardt's solution, 50% dextran sulfate). Approximately 782 μl of this solution is prepared, using 500 μl formamide, 60 μl 5 *M* NaCl, 10 μl 1 *M* Tris pH 8.0, 2 μl 0.5 *M* EDTA pH 8.0, 10 μl 100X Denhardt's solution, and 200 μl 50% dextran sulfate. This solution should be thoroughly mixed and may be stored at 4°C for several weeks.
2. **Preparation of Hybridization Solution B** (tRNA 11 mg/ml, RNA probe, 1 *M* DTT, distilled water). Approximately 218 μl of this solution is prepared, using 50 μl tRNA 11 mg/ml, the RNA probe, 10 μl 1 *M* DTT, and distilled water to bring the volume up to 218 μl. The solution is mixed and heated for 5 min at 65°C. This solution may be stored at –20°C.
3. **Combined Hybridization Solution.** Solutions A and B should be combined and mixed, and then centrifuged at 5000 rpm for 10 min at room temperature. If it is stored, it must be placed in a 65°C water bath for 10 min to melt the annealed strands before use.
4. **Hybridization of Tissue Sections.** Approximately 75 μl of the combined hybridization solution is placed on a 22 × 60 mm coverslip, so that the solution flows over the tissue without bubbles. Wipe off excess solution and seal the edges of the coverslip to the slide with liquid DPX. The slides are incubated overnight on a slide warming tray at 55°C. The tray may be protected from the radioactive material with aluminum foil.

H. Posthybridization

The slides are rinsed in a buffer solution. RNase is used to digest all nonspecific, unbound or bound ssRNA. High stringency washes are then used to remove non-specifically bound probe and to stabilize the hybridized RNAs.

1. **Removal of Hybridization Solution** (4X SSC). The slides are removed from the warming tray and are allowed to cool at room temperature. The dried DPX is peeled

off the slides with sharp forceps. The coverslips are removed with gentle agitation on a rotator table in a 4X SSC rinse for 15 min. After the coverslips are removed, an additional four 5-min rinses in 4X SSC should be performed on a rotator table set for gentle agitation. The level of radioactivity of these steps is high until the last rinse.

2. **RNase Digestion** (10 mg/ml RNase A, 5 *M* NaCl, 1 *M* Tris pH 8.0, 0.5 *M* EDTA pH 8.0, distilled water). The slides are immersed in an RNase solution for 30 min at 37°C. To prepare a 300-ml solution, 600 µl 10 mg/ml RNase A, 30 ml 5 *M* NaCl, 3 ml 1 *M* Tris pH 8.0, 600 µl 0.5 *M* EDTA pH 8.0, and 266 ml distilled water are combined. The level of radioactivity of this step is high.
3. **High-Stringency Washes** (2X SSC, 0.1X SSC, 1 m*M* DTT). The slides are placed in 300 ml of 2X SSC plus 300 µl 1 *M* DTT for 30 min at room temperature. The level of radioactivity of this step is low. The slides are placed in 0.1X SSC plus 300 µl 1 *M* DTT for 30 min at 55°C. The level of radioactivity of this step is high. The slides are then briefly rinsed in 0.1X SSC plus 300 µl 1 *M* DTT for 5 min at room temperature. The level of radioactivity of this step is low.
4. **Dehydration** (50%, 70%, 95%, 100% ethanol; 20X SSC). The slides are dehydrated using the following series of solutions: 50% ethanol with 1.5 ml 20X SSC for 3 min, 70% ethanol with 1.5 ml 20X SSC for 3 min, 95% ethanol for 3 min, and 3 changes of 100% ethanol for 3 min each. The slide racks should be drained and vacuum dried at room temperature for 30 min.

I. Autoradiography

The slides and ^3H standards are exposed to film, which is subsequently developed. Densitometry is used to analyze the film.

1. **Film Exposure.** The slides and ^3H standards are taped to cardboard and microautoradiography emulsion (e.g., Reichert-Jung Ultrofilm) and placed in film cassettes for a 30-day period at 4°C.
2. **Optical Densitometry and Controls.** The ISHH signal in the brain sections is quantified using a commercial image processing and analysis program. The use of the ^3H standards on the film provide standardization of the optical density measurements. Sense strand hybridization under identical conditions may be used as a control; a minimal, homogeneous signal should be obtained when the riboprobe is hybridized in this control situation. Other possible controls[5] include: 1) oligonucleotides complementary to different regions of the receptor subtype should provide similar binding; 2) competition: specific hybridization signals disappear with an excess of unlabeled oligonucleotide in the hybridization solution, but no alteration of the signal is present when the unlabeled oligonucleotide added in excess is complementary to a different region of the same mRNA; 3) specificity of signal found by different hybridization patterns with probes for related but different mRNAs; 4) thermal stability: hybridization signal intensity changes observed at a temperature consistent with the theoretical T_m of the hybrids; and 5) receptor autoradiography and immunohistochemistry if selective ligands of specific antibodies exist for a given receptor. This protocol is summarized in Table 10.1.

TABLE 10.1
Summary of Protocol

- **Perfusion:** pH shift perfusion with a 4% paraformaldehyde solution.
- **Slide Preparation:** Slides are coated with a solution of gelatin, chrome alum, and poly L-lysine, which promote tissue adherence and also decrease nonspecific probe binding.
- **Tissue Sectioning:** Following removal, blocking, and freezing, the brain is sectioned in a cryostat and the sections are placed directly on the prepared slides.
- **Tissue Permeabilization:** With fixed tissue, the mRNA may be unavailable to the probe; thus, tissue deproteination by proteinase K increases probe access.
- **Prehybridization:** The tissue is placed in a buffer solution, acetylated to block the positive charges from the lysine, rinsed, and then dehydrated with ethanol.
- **Riboprobe Preparation:** Using the RNase-free, linearized DNA as a template, a complementary radioactively labeled RNA transcript is obtained. The specific activity of the probe is obtained by a scintillation counter. The probe is purified by removing the DNA template with RNase-free DNase I. The probe is then extracted with phenol and chloroform; ethanol precipitation and size-exclusion chromatography are used to remove the majority of unincorporated nucleotides and to concentrate the RNA.
- **Hybridization:** The RNA probe solution is prepared and applied to the tissue and allowed to incubate overnight.
- **Posthybridization:** The slides are rinsed in a buffer solution. RNase is used to digest all nonspecific, unbound or bound ssRNA. High stringency washes are then used to remove nonspecifically bound probe and to stabilize the hybridized RNAs.
- **Autoradiography:** The slides and ^{3}H standards are exposed to film, which is subsequently developed. Densitometry is used to analyze the film.

III. Conclusion

We used *in situ* hybridization histochemistry to study REM-sleep deprivation effects on m1–m3 muscarinic receptor mRNA expression in the rat brain. REM-sleep deprivation for 72 hours did not affect m1 receptor mRNA expression. However, we found significantly increased m3 receptor mRNA expression in the pontine nuclei and nucleus accumbens-bed nucleus of the stria terminalis region of REM-sleep deprived rats versus controls. Paradoxically, we found significantly decreased m2 receptor mRNA expression in the pontine nuclei of REM-sleep deprived rats versus controls. The present findings implicate these rat brain structures in the cholinergic effector pathways of REM sleep. However, the type and magnitude of the effects of these structures on REM sleep varied with different receptor subtypes; further research is necessary to delineate the specific functions and the extent of involvement these structures provide in REM sleep mechanisms.

An example of the m1 muscarinic receptor mRNA autoradiograms obtained from brain sections of the caudate/putamen and the hippocampus/dentate gyrus regions of a REM-sleep deprived rat and control rat pair are depicted in Figure 10.1. For these brain structures, no significant differences were detected in the m1 ISHH signal of REM-sleep deprived rats compared to controls.

No technique is perfect; there are limitations to ISHH, which include: 1) a change in overall gene expression (i.e., as measured by oligo(dT) ISHH for detection of poly (A) tails of mRNA) may mask an opposite change in specific receptor subtype

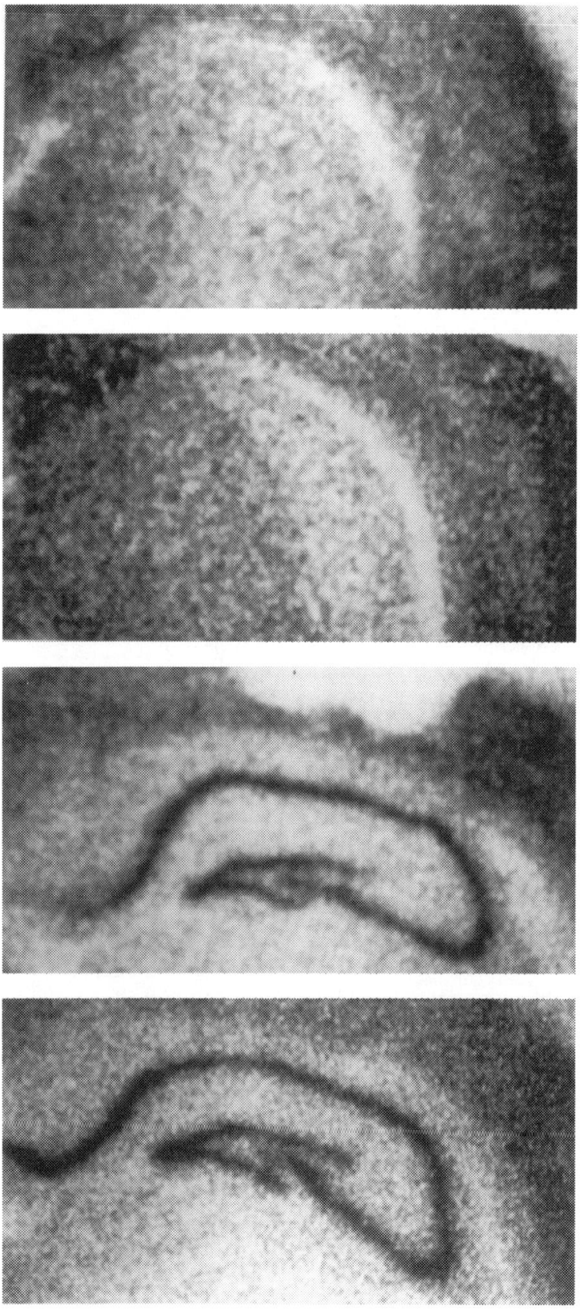

FIGURE 10.1
Photographs of the m1 muscarinic receptor mRNA autoradiograms obtained from brain sections of the caudate/putamen region of a REM-sleep-deprived and control rat pair (top two photographs, respectively) and the hippocampus/dentate gyrus region for a REM-sleep-deprived and control rat pair (bottom two photographs, respectively). No significant differences were detected in the optical density in these brain structures for m1 ISHH signal in REM-sleep-deprived rats compared to controls.

expression;[15] 2) small, discrete nuclei may be below the resolution threshold of ISHH; 3) exact quantification of transcripts in terms of number of mRNA molecules per cell is difficult, but relative quantitative estimates are obtainable by microdensitometric measures;[5] 4) changes in muscarinic receptor mRNA may not be accompanied by an equivalent change in receptor protein, since changes in translational efficiency or post-translational mechanisms (glycosylation or phosphorylation) may occur with the experimental manipulation;[16] 5) change in mRNA expression may not be associated with concomitant change in ligand binding because mRNA expression occurs primarily in cell bodies, while ligand binding occurs on both dendrites and axon terminals;[16] and 6) there is no consideration of the functional properties of the receptors (i.e., G-protein activation, phosphoinositol turnover) which may be affected by the experimental condition.[17] Despite these limitations, ISHH serves as a useful and flexible technique to identify cells synthesizing receptors, and offers advantages over conventional radioligand autoradiography and immunohistochemistry.

References

1. Simmons, D. M., Arriza, J. L., and Swanson, L. W., A complete protocol for *in situ* hybridization of messenger RNAs in brain and other tissues with radio-labeled single-stranded RNA probes, *J. Histotechnol.,* 12(3), 169, 1989.

1a. Conn, P. M., Ed., *Methods in Neurosciences*, Academic Press, San Diego, 1988, 79.

2. Mash, D. C. and Potter, L. T., Autoradiographic localization of M1 and M2 muscarine receptors in the rat brain, *Neuroscience*, 19(2), 551, 1986.

3. Harrison, P. J., Barton, A. J. L., Najlerahim, A., McDonald, B., and Pearson, R. C. A., Increased muscarinic receptor messenger RNA in alzheimer's disease temporal cortex demonstrated by in situ hybridization histochemistry, *Mol. Brain Res.*, 9, 15, 1991.

4. Blake, M. J., Appel, N. M., Joseph, J. A., Stagg, C. A., Anson, M., DeSouza, E. B., and Roth, G. S., Muscarinic acetylcholine receptor subtype mRNA expression and ligand binding in the aged rat forebrain, *Neurobiology of Aging*, 12, 193, 1991.

5. Palacios, J. M., Mengod, G., Sarasa, M., Vilaró, M. T., Pompeiano, M., and Martinez-Mir, M. I., The use of *in situ* hybridization histochemistry for the analysis of neurotransmitter receptor expression at the microscopic level, *J. Recept. Res.*, 11 (1–4), 459, 1991.

6. Kushida, C., Zoltoski, R. K., and Gillin, J. C., The expression of m1–m3 muscarinic receptor mRNAs in rat brain following REM sleep deprivation, *NeuroReport*, 6, 1705, 1995.

7. Velazquez-Moctezuma, J., Shalauta, M., Gillin, J. C., and Shiromani, P. J., Cholinergic antagonists and REM sleep generation, *Brain Res.*, 543, 175, 1991.

8. de Jonge, A., Doods, H. N., Riesbos, J., and van Zwieten, P. A., Heterogeneity of muscarinic binding sites in rat brain, submandibular gland and atrium, *Br. J. Pharmac.*, 89, Suppl., 551P, 1986.

9. Michel, A. D., Stefanich, E., and Whiting, R. L., PC12 phaeochromocytoma cells contain an atypical muscarinic receptor binding site, *Br. J. Pharmac.*, 97, 914, 1989.

10. Bonner, T. I., The molecular basis of muscarinic receptor diversity, *Trends Neurosci.*, 12, 148, 1989.

11. Buckley, N. J., Bonner, T. I., Buckley, C. M., and Brann, M. R., Antagonist binding properties of five cloned muscarinic receptors expressed in CHO-K1 cells, *Molec. Pharmac.*, 35, 469, 1989.
12. Fukuda, K., Higashida, H., Kubo, T., Maeda, A., Akiba, I., Bujo, H., Mishina, M., and Numa, S., Selective coupling with K^+ currents of muscarinic acetylcholine receptor subtypes in NG108-15 cells, *Nature*, 335, 355, 1988.
13. Bonner, T. I., Buckley, N. J., Young, A. C., and Brann, M. R., Identification of a family of muscarinic acetylcholine receptor genes, *Science*, 237, 527, 1987.
14. Berod, A., Hartman, B. K., and Pujol, J. F., Importance of fixation in immunohistochemistry: use of formaldehyde solutions at variable pH for the localization of tyrosine hydroxylase, *J. Histochem. Cytochem.*, 29(7), 844, 1981.
15. Harrison, P. J., Procter, A. W., Barton, A. J. L., Lowe, S. L., Najlerahim, A., Bertolucci, P. H. F., Bowen, D. M., and Pearson, R. C. A., Terminal coma affects messenger RNA detection in post mortem human temporal cortex, *Mol. Brain Res.*, 9, 161, 1991.
16. Harrison, P. J., Barton, A. J. L., Najlerahim, A., McDonald, B., and Pearson, R. C. A., Increased muscarinic receptor messenger RNA in Alzheimer's Disease temporal cortex demonstrated by in situ hybridization histochemistry, *Mol. Brain Res.*, 9, 15, 1991.
17. Blake, M. J., Appel, N. M., Joseph, J. A., Stagg, C. A., Anson, M., DeSouza, E. B., and Roth, G. S., Muscarinic acetylcholine receptor subtype mRNA expression and ligand binding in the aged rat forebrain, *Neurobiology of Aging*, 12, 193, 1991.

Chapter 11

Transcriptional Regulation of Putative Sleep-Promoting Compounds

Miroslaw Mackiewicz, Sigrid C. Veasey, and Allan I. Pack

Contents

I. Introduction

Current models of control of sleep and wakefulness posit that during wakefulness there is a growing pressure for sleep (process S) that is dissipated during sleep.[1] It is assumed that this process is neurochemically mediated, although its precise molecular nature is unknown. There is, however, no shortage of candidate molecules, and

0-8493-3361-X/98/$0.00+$.50

there are some data about the role of a large variety of neurochemicals of different types, e.g., hormones, peptides, cytokines, prostaglandins, etc., as sleep-promoting compounds and as controls of sleep and wakefulness.[2] None of the molecules so identified is thought to be specific to sleep. This is unlike studies in the circadian system where genes specific to the molecular mechanism of the clock have been identified in *Drosophila* (reviewed in reference 3; see also references 4 and 5) and in neurospora.[6] This lack of specificity will have an impact on the molecular and genetic approaches that are likely to prove fruitful in studies of the basic regulatory mechanisms of sleep.

Some of the putative sleep-promoting compounds, e.g., interleukin-1β (IL-1β), are produced by action of a single gene, while the levels of other compounds, such as adenosine, are determined by a complex set of interrelated enzymes,[7] all of which can be transcriptionally regulated. Thus, no single paradigm is likely to prove applicable to all putative sleep-promoting compounds. There is, moreover, no guarantee that levels of all putative sleep-promoting compounds will have the same relationship to the duration of prior wakefulness. Indeed, intuitively this seems unlikely. Some molecules may show regulation with the normal sleep–wake cycle. Others may only be activated after prolonged sleep deprivation and thus be more in the nature of back-up systems. It seems likely that as this area of investigation develops, a sleep cascade will emerge as it has for other fundamental biological functions. All of these issues need to be considered when proposing studies of the molecular basis for the regulation of sleep and wakefulness.

A beginning point for many of these studies will be to assess whether concentrations of the molecule in question are regulated in specific brain regions thought to be relevant for sleep control. The primary purpose of such studies would be to determine whether levels of specific proteins change in relation to sleep and wakefulness and following sleep deprivation. Since overall cytosolic protein levels change across the circadian day,[8] results of such studies need to be interpreted with caution. Moreover, studies of putative sleep-promoting compounds need to utilize behavioral paradigms that allow one to separate regulation of molecules that are coupled to the circadian system from those linked to the sleep homeostatic process. Thus, studies of changes in protein level may be complemented by studies involving assessment of changes in specific mRNAs. For this, a variety of techniques are available but the relatively low abundance of many of the relevant mRNAs limits the approaches that can be used. (This is discussed more fully below.) Studies to quantify mRNA levels need to be complemented by techniques such as *in situ* hybridization[9] to localize where and in what cells identified changes in mRNA occur.[10]

II. Methods of mRNA Analysis

The regulation of gene expression is accomplished in part by diversity and flexibility of mechanisms that control gene transcription. Changes in transcription level cause variation in the level of individual molecules of messenger RNA (mRNA). Thus, the analysis of mRNA is of prime importance in understanding mechanisms of gene regulation. Traditionally, the estimation of mRNA level and its changes have been accomplished by the DNA/RNA or RNA/RNA hybridization methods. Northern blot and nuclease protection assays have been used most extensively.

In the Northern blot, the size and amount of mRNA are determined autoradiographically following gel electrophoresis, transfer to a solid support (nitrocellulose or nylon membrane), and hybridization with a radiolabeled DNA or RNA probe.[11,12] The ability to determine the size of the mRNA of interest is a major advantage in Northern blot hybridization. This method requires, however, a large amount of total RNA, in excess of several micrograms even when the gene is expressed at high levels. Sensitivity of the Northern blot can be increased by further purification of total RNA by oligo(dT) cellulose chromatography (see reference 12), but the procedure is time consuming and requires a considerable amount of starting material, i.e., total RNA.

In the ribonuclease (RNase) protection assay, a hybridization between the radiolabeled probe and target RNA take place in solution rather than on the solid support. After hybridization, the remaining probe RNA and the target RNA are degraded by ribonucleases, and products are separated electrophoretically, followed by autoradiography.[13] Nuclease protection assays are at least ten times more sensitive than Northern blot hybridization and are particularly useful when detection and quantitation of more than one mRNA species is desirable.[14] Since multiple probes can be used simultaneously, an internal standard can be included in the same reaction. Unlike Northern blot hybridization, the RNase protection assay requires a minute amount of target RNA and tolerates partially degraded RNA. The nuclease protection assay can distinguish between mRNAs of multigene families which comigrate or cross-hybridize on the Northern blots. In addition, the assay can be used to map transcription initiation and termination sites or intron–exon boundaries.[12]

Application of polymerase chain reaction (PCR) has provided a major breakthrough in the approach to the analysis of gene transcription.[15] Its main advantage is simplicity and a sensitivity that is several orders of magnitude higher than that of Northern blot hybridization or of RNase protection assays. It allows the detection of small quantities of specific mRNAs, and thus may be used to analyze gene transcripts of single cell. PCR is an enzymatic process of amplification of a template using primers designated to hybridize to opposing strands of target DNA. The PCR-based method of analysis of mRNA is termed reverse transcription–polymerase chain reaction (RT–PCR). This is a two-step process that requires reverse transcription, i.e., synthesis of cDNA (DNA complementary to RNA), followed by polymerase chain reaction. RT–PCR products are verified based on their sizes following gel electrophoresis. In addition, products of polymerase chain reaction can be further validated by restriction enzyme digestion, nucleotide sequencing and/or Southern blot hybridization. Quantitation of RT–PCR products can be achieved by measuring the incorporation of radioactively labeled nucleotide or primers into DNA. Alternatively, an autoradiogram of the gel can be generated and analyzed densitometrically. The various methods of quantitation of PCR products are reviewed in reference 16.

III. Quantitative Analysis of mRNA by RT–PCR

RT–PCR is a very sensitive and straightforward method of mRNA analysis; obtaining quantitative data with this technique is, however, difficult. RT–PCR is a two-step process, i.e., reverse transcription and polymerase chain reaction. Thus, both steps must be considered in quantitation of mRNA. In practice, however, the second step (PCR) poses the more serious barrier in acquiring quantitative data. Because of the exponential character of the PCR, small changes in amplification efficiency produce considerable differences in the reaction yield. For instance, over the course of 30 cycles of amplification, a 5% difference in amplification efficiency produces more than a twofold change in the yield of PCR products. In addition, quantitation of mRNA levels by RT–PCR is limited to analysis of products generated only in the exponential phase of amplification, during which its concentration is proportional to the concentration of the starting target cDNA. At the later stages of amplification, the reaction slows down substantially and reaches a plateau. Thus, the plateau effect contributes also to the difficulty of performing quantitative PCR.

Accurate quantitation of mRNA requires correction for reaction-to-reaction variation in the yield of amplification. However, if the amplification efficiencies are the same and the PCR products are collected before the reaction reaches its plateau, the mRNA level, or more precisely, the relative difference in target mRNA can be determined by titration[17] or kinetic analysis.[18] Using linear regression methodology proposed by Wiesner,[19] it is possible to estimate the absolute number of mRNA molecules per unit of total RNA and to calculate the number of molecules per cell.

Difficulties in estimation of amplification efficiency can be addressed by including an internal standard in RT–PCR. Endogenous gene transcripts or exogenous fragments can be used in quantitative RT–PCR. Endogenous standards are usually mRNA for housekeeping genes or mRNA for structural elements of the cell that are present in the RNA sample used for RT–PCR. Transcripts of the gene that are structurally or functionally related to the target mRNA have also been used for quantitative analysis.[20-24] Following the titration or kinetic analysis of PCR, the ratio of PCR product for an endogenous standard and a target molecule are compared between samples, and by using linear regression analysis,[19] the relative initial amounts of target and standard can be determined. It has been demonstrated that, if an internal standard is expressed at the same level in two samples, the ratio of PCR product for a standard and a target indicates the relative level of target mRNA.[20] Thus, the linear regression analysis of absolute amounts of the target and standard can be omitted. RT–PCR with an endogenous standard has been used extensively to determine the relative level and the changes of specific transcripts.[20-22,25] When amplification of an endogenous mRNA is used to standardize RT–PCR, several factors can complicate the interpretation of the results. This approach depends on the assumption that the level of expression of a reference standard is the same in each sample and does not change during the experimental manipulation. However, few if any genes have a strict constitutive pattern of expression, including the most commonly used reference transcript, e. g., mRNA for β-actin. Also, it is important to determine the PCR kinetics for a standard and target to avoid the plateau phase, especially when the relative levels of both molecules differ substantially (e.g., most cases when β-actin is used as a standard). Interference of primers for a standard and target during PCR is a common problem in amplification experiments using an endogenous sequence as a reference. This leads to a reduction in amplification efficiency and premature termination of the exponential phase.

RT–PCR and/or PCR can also be standardized by amplification of an exogenous sequence as a reference. This standard is usually a synthetic RNA added directly to the reaction of reverse transcription or DNA that is included only in PCR. Quantitation of mRNA from RT–PCR with an exogenous standard is similar to that described above for the endogenous reference transcripts. The advantage of using an exogenous sequence as a reference is that the initial amount of the standard is known and can be used to determine the absolute level of target mRNA. In the method described by Wang et al.[26] a synthetic internal standard with the binding sites for primers, identical to those in the target but designated to generate PCR products of different sizes, is included in the reverse transcription reaction. Following the serial dilution of cDNA, PCR titration curves for a standard and target are generated. The number of target mRNA molecules per unit of total RNA is determined by plotting an experimental and standard curve and is then compared in the region where they parallel each other. (See reference 26 for a detailed description of the procedure for plotting the curves and the determination of the number of target molecules.) Since the reference and target molecules are not titrated against each other, the reaction is not considered to be a competitive PCR. Since the report by Wang et al.[26] appeared, exogenous RNA or DNA have been used as a standard for many studies.[27-30]

In competitive PCR, a series of dilutions of an exogenous standard, which is a different size than the target, are added to a series of PCR reactions that contain a constant amount of target molecules (or vice versa).[31-32] It is assumed that the molar ratio of a standard and target is equal in those reactions where the yield of PCR products for a standard and target are the same. Since the amount of competitor added to the reaction of reverse transcription and PCR is known, the absolute initial amount of the target can be determined. The important advantage of competitive PCR is that quantitative data can be obtained even after the reaction has reached the plateau phase of amplification. In experiments where the relative changes in mRNA level are to be determined, it is not, however, necessary to perform a full dilution series of the competitor.[32] A quantitative analysis can be performed by adding a constant amount of a competitor to the PCR and estimating the level of target among samples by comparing the ratio of the amplified target to the amplified standard in each sample. The differences in this ratio between the samples reflect the level of target in each sample. This approach is especially useful when studying multiple samples. In competitive PCR, a standard competitor can be either a heterologous fragment with identical primer sequences[34,35] or a homologous fragment yielding PCR product of sizes different from that of the target.[28,29]

We have used a competitive PCR with a constant amount of homologous competitor fragment to determine the relative changes in the level of mRNA for IL-1β in the rat brain following an experimental sleep manipulation. The details of the experimental procedures and results are described below.

IV. Changes in IL-1β Gene Expression in the Rat CNS During the Sleep–Wake Cycle and Following Sleep Deprivation — An Example of the Use of RT–PCR

There is considerable data that IL-1β can modify sleep and that it may play a role in the control of sleep. [36] For example, administration of IL-1β into the ventricles increases the amount of slow-wave sleep in rabbits[37] and rats[38]; in the latter, this effect is dependent on circadian phase.[38] Likewise, administration of the IL-1 receptor antagonist[39] or anti-IL-1β antibody[40-41] has minor effect on normal normal sleep–wake behavior, but has significant effect on recovery sleep following a period of sleep deprivation.[39-42] Neurons expressing IL-1β have been identified in the hypothalamus,[43] and there are receptors for IL-1β in relevant brain regions, with respect to sleep control.[44] Thus, there is good reason to believe that IL-1β could play a role in normal sleep–wake control and particularly in response to sleep deprivation.

One important test of whether IL-1β plays such a role is whether its level in relevant brain regions is modulated in relation to the sleep–wake cycle and in response to sleep deprivation. Since it is present in low abundance, RT–PCR as we have described, is one approach to address this question.

Adult, male Sprague-Dawley rats were used in our experiments. Under general anesthesia EEG electrodes were implanted in the skull and EMG electrodes were inserted in the dorsal neck musculature. Power density spectra of EEG were computed online using the data analysis software ASYST (ASYST Software Technologies Inc.). The integrated power of the EEG along with the mean value of the averaged EMG provided a training set for an artificial neural network used to detect sleep. Sleep deprivation was achieved by periodic forced locomotion by rotation of the floor of the chamber holding the animals. A yoked control animal experienced an equal amount of forced locomotion, but was able to sleep whenever the chamber floor was stationary. The experimental animal was awakened as soon as the computer system detected sleep. This is a modification of the paradigm of Rechtschaffen et al.[45] In a separate set of experiments, wheel-running activity was monitored by means of a computer-aided data acquisition system (Chronobiology Kit, Stanford Software Systems). Based upon wheel-running data, animals were decapitated under dim red illumination at either circadian time (CT) 10 or 22 (two hours before the preferred activity-phase or rest-phase, respectively).

For RNA isolation the brain was removed and dissected into four regions: cerebral cortex, thalamus, hypothalamus, and the brain stem. RNA was isolated by the modified guanidinium thiocyanate method of Chomczynski and Sacchi.[46]

In our studies, a fragment of the rat IL-1β gene was amplified using RNA isolated from rat brain as described by Minami et al.[47] and cloned using the TA cloning kit (Invitrogen). Verification of RT–PCR products was performed both by restriction enzyme digestion and nucleotide sequencing using the dideoxy chain termination method.[12] An IL-1β gene mutant carrying a seventeen-base-pair deletion between nucleotides 455 and 472 of the coding region was created by Pfl MI restriction enzyme digestion following limited degradation using the exonucleolitic activity of

T4 DNA polymerase.[12] Plasmid templates for *in vitro* transcription were purified by Qiagen columns and linearized by restriction enzyme digestion. The standard for RT–PCR was prepared by *in vitro* transcription of the mutated gene using SP6 RNA polymerase.[12] Prior to cDNA synthesis, the DNA template was removed by extensive DNase I digestion. For quantitation of IL-1β mRNA, the amount of standard, as determined by RT–PCR, was diluted so that its level approximated the level of IL-1β in the brain. cDNA was synthesized by random priming, using 5 μg of total RNA, internal standard RNA, 0.5 μg of DNA random hexanucleotides, and 200 units of M-MLV reverse transcriptase. Aliquots (5 μl) of the RT reaction were subjected to PCR using 1 unit of Taq DNA polymerase (Perkin Elmer) in the presence of 5 μCi of α-^{32}P-dCTP (3000 Ci/mmol). Primers (100 ng each) used in our experiments were: 5′–gtgggatgatgacgacctgc–3 (sense)′, and 5′–ttgtcgttgcttgtctctcc–3′ (antisense). Twenty nine PCR cycles (94°C for 30 sec, 58°C for 1 min and 72°C for 30 sec) were performed in an automated thermal cycler (Perkin Elmer). 5 μl aliquots of the PCR mixture were electrophoresed on 6% polyacrylamide gel; dried gels were exposed to X-ray films (X-Omat, Kodak) for 6 to 12 h at room temperature. The X-ray films were quantified using a computer-assisted laser scanner (Ultroscan XL, LKB) and analyzed by 2400 GelScan XL software). The amount of mRNA for IL-1β was expressed as a ratio of densitometric measurements of the RT–PCR product derived from the IL-1β target sequence and the competitive standard.

As the result of RT–PCR, two bands were visible on the gel after electrophoresis: the 168 base pairs standard IL-1β and the 185 base pairs fragment of IL-1β gene originating from the cytokine mRNA present in the brain (Figure 11.1). In order to determine the effect of surgical procedures on transcription of the IL-1β gene, estimation of mRNA for this cytokine was performed using RNA from the CNS of normal and sham operated rats in which the EEG and EMG electrodes were implanted but no sleep deprivation was attempted. Two weeks after surgery, the quantity of IL-1β mRNA was similar in normal and sham operated rats (data not shown). The level of mRNA for IL-1β in the cerebral cortex, thalamus, hypothalamus, and brain stem of normal control rats, yoked controls, and rats subjected to 24 h of sleep deprivation is presented in Figure 11.2. By one-way analysis of variance (ANOVA) there were significant differences in the level of mRNA for IL-1β in the

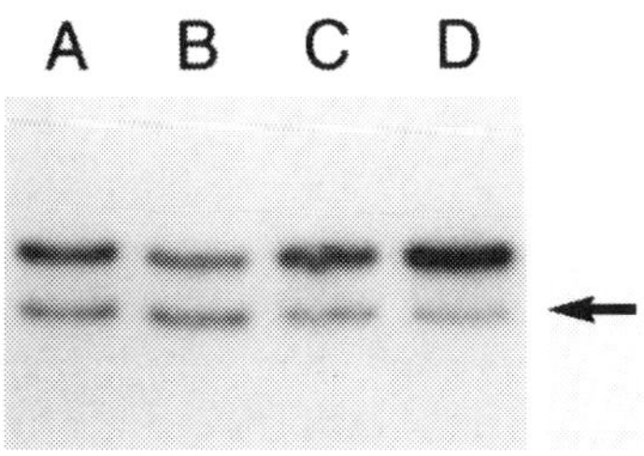

FIGURE 11.1
Detection of mRNA for IL-1β in the rat hypothalamus by RT–PCR. The lines A to D correspond to normal, sham operated, yoked control, and sleep-deprived rats, respectively. The arrow indicates the IL-1β standard (From Mackiewicz, M., Sollars, P. J., Ogilvie, M. D., and Pack, A. I., Modulation of IL-1β gene expression in rat CNS during sleep deprivation, *NeuroReport,* 7, 529, 1996. With permission.)

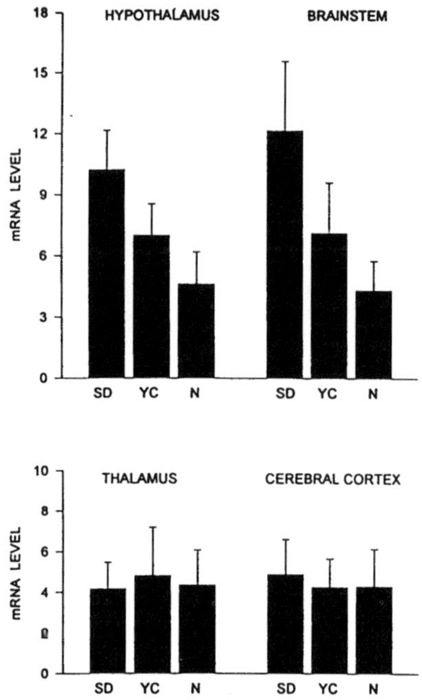

FIGURE 11.2
The level of mRNA for IL-1β in the CNS of normal (N), yoked control (YC), and sleep-deprived (SD) rats as determined by RT–PCR. The amount of mRNA for IL-1β is presented in the arbitrary units obtained as the ratio of densitometric measurements of the samples and corresponding standard. Note the marked increase in the level of mRNA for IL-1β in the hypothalamus and the brain stem of rats subjected to sleep deprivation (From Mackiewicz, M., Sollars, P. J., Ogilvie, M. D., and Pack, A. I., Modulation of IL-1β gene expression in rat CNS during sleep deprivation, *NeuroReport,* 7, 529, 1996. With permission.)

hypothalamus and brain stem ($p < 0.0002$; $p < 0.0001$, respectively) between normal control (baseline) animals, yoked controls, and the rats subjected to 24 h of sleep deprivation. The differences between different brain regions were also observed in yoked control animals ($p < 0.043$) but not in normal rats. An unpaired *t* test between values from different regions of the brain of animals subjected to sleep deprivation and corresponding yoked control rats revealed significant differences in the level of mRNA for IL-1β in the hypothalamus and the brain stem regions ($p < 0.009$ and $p < 0.014$, respectively). However, by unpaired *t* test there were no differences in the amount of messenger RNA for this cytokine in the thalamus or cerebral cortex of sleep-deprived as compared to corresponding yoked control rats. Finally, to determine whether the results obtained following sleep deprivation would be reflected similarly in changes of IL-1β mRNA during different phases of the normal circadian cycle, the level of transcript was analyzed at CT 10 (two hours before the projected onset of activity) and CT 22 (two hours before the projected onset of the preferred rest period). The circadian rhythmicity of wheel-running activity in constant environmental conditions was used as a phase marker. There were no detectable differences in the level of mRNA for IL-1β at these two circadian time points in any brain region tested (data not shown).

V. Discussion

Traditional methods for the detection and analysis of gene transcripts require substantial amounts of RNA even when the gene is expressed at high levels. In some experimental models, including research on the gene expression in the CNS, transcription of many genes is at low levels and/or only in a specific group of cells. In such cases, RT–PCR has proven to be an invaluable tool. Since the PCR is an exponential process, the variation in amplification efficiency yields large differences in the amount of products. In addition, at later stages of PCR, the amplification rate slows down and ultimately levels off.[48] Both characteristics of PCR complicate its reliability as a quantitative method. However, if certain conditions and controls are applied, quantitative information about mRNA and its changes can be obtained. Of the various quantitative RT–PCR techniques, competitive PCR seems to be the most reliable and circumvents many disadvantages of the other quantitative or semiquantitative approaches.

Due to the low levels of mRNA for IL-1β in the brain, RT–PCR was the method of choice in our experiments. To quantify the amount of IL-1β mRNA, an endogenous standard RNA (IL-1β deletion mutant) designed to generate a PCR product of a different size was included in the cDNA synthesis reaction and successfully used in competitive RT–PCR. cDNA synthesis by random priming was the method of choice, because RNA used as an internal standard was devoid of polyadenylated sequence. It is generally believed, that oligo(dT) priming is more efficient than priming with random hexamers. However, when these two methods were compared in our laboratory by measuring the level of mRNA for β-actin, there were no major differences in the efficiency of the reaction as well as the quality of RT–PCR products.

In our experiments, we have demonstrated that there is a marked increase in the level of mRNA for IL-1β in the hypothalamus and brain stem of rats after 24 h of sleep deprivation. The results presented here support the hypothesis that IL-1β is involved in the homeostatic component of sleep regulation and that the transcription of its gene increases during sleep deprivation. Hence, IL-1β may have a regulatory role in normal, and especially recovery, sleep following sleep deprivation. The factors and mechanisms regulating the central expression of IL-1β remain to be determined.

This example shows that RT–PCR can be used to study changes in transcriptional regulation of the genes that might be involved in sleep control. This approach is, however, only one component of an overall strategy to investigate the role of specific molecules in sleep control. Other components involve studies of how a specific gene is regulated; in what cell types this regulation is taking place; what effects this molecule has at a cellular or molecular level, e.g., does the molecule act to induce expression of other genes involved in neuronal signaling; whether alteration of the expression of this gene in transgenics with overexpression of the gene or loss of the gene in knockouts alters the behavior of interest, i.e., sleep. We believe that bringing all of these paradigms to bear in a complementary way will yield a fundamental understanding of the mechanisms controlling sleep.

Acknowledgments

This work was supported by NIH grants HL-07713 and HL-42236 and the American Sleep Apnea Association Award. The studies describing changes in mRNA for IL-1β with sleep deprivation were conducted in collaboration with Dr. Patricia J. Sollars and Dr. Malcolm D. Ogilvie.

References

1. Borbely, A. A., A two process model of sleep regulation, *Hum. Neurobiol.*, 1, 195, 1982.
2. Borbely, A. A. and Tobler, I., Endogenous sleep-promoting substance and sleep regulation, *Physiol. Rev.*, 69, 605, 1989.
3. Rosbash, M. and Hall J.C., The molecular biology of circadian rhythms, *Neuron,* 3, 387, 1989.
4. Myers, M. P., Wager-Smith, K., Wesley, C. S., Young, M. W., and Sehgal, A., Positional cloning and sequence analysis of the *Drosophila* clock gene, *timeless, Science*, 270, 805, 1995.
5. Sehgal, A., Rothenfluh-Hilfiker, A., Hunter-Ensor, M., Chen, Y., Myers, M. P., and Young, M. W., Rhythmic expression of *timeless*: a basis for promoting circadian cycles in *period* gene autoregulation, *Science,* 270, 808, 1995.
6. McClung, C. R., Fox, B. A., and Dunlap, J. C., The *Neurospora* clock gene *frequency* shares a sequence element with the *Drosophila* clock gene *period, Nature,* 339, 558, 1989.
7. Phillis, J.W., *Adenosine and Adenine Nucleotides as Regulators of Cellular Function,* CRC Press, Boca Raton, 1991.
8. Ayers, N. A., Kapas, L., and Krueger, J. M., Circadian variation of nitric oxide synthase activity and cytosolic protein levels in rat brain, *Brain Res.,* 707, 127, 1996.
9. Shivers, B. D., Schachter, B. S., and Pfaff, D. W., *In situ* hybridization for a study of gene expression in the brain, *Methods Enzymol.,* 124, 497, 1986.
10. Breder, C. D. and Saper, C.D., Expression of inducible cyclooxygenase mRNA in the mouse brain after systemic administration of bacterial lipopolysaccharide, *Brain Res.,* 713, 64, 1996.
11. Alwine, J. C., Kemp, D. J., and Stark, G. R., Method for detection of specific RNAs in agarose gel by transfer to diazobenzyloxymethyl-paper and hybridization with DNA probes, *Proc. Natl. Acad. Sci. USA,* 74, 5350, 1977.
12. Sambrook, J., Fritsch, E. F., and Maniatis, T., *Molecular Cloning: a laboratory manual.* Cold Spring Harbor Laboratory Press, Cold Spring Harbor, NY, 1989.

13. Berk, A. J. and Sharp, P. A., Sizing and mapping of early adenovirus mRNAs by gel electrophoresis of S1 endonuclease digested hybrids, *Cell*, 12, 721, 1977.
14. Lee, J. J. and Costlow, N. A., A molecular titration assay to measure transcript prevalence levels, *Methods Enzymol.*, 152, 633, 1987.
15. Veres, G., Gibbs, R. A., Scherer, S. E., and Caskey, C. T., The molecular basis of sparse fur mouse mutation, *Science*, 237, 415, 1987.
16. Brenner, C. A., Tam, A. W., Nelson, P. A., Engleman, E. G., Suzuki, N., Fry, K. E., and Larrick, J. W. Message amplification phenotyping (MAPPing): A technique to simultaneously measure multiple mRNA from small numbers of cells, *Biotechniques*, 7, 1096, 1989.
17. Singer-Sam, J., Robinson, M. O., Belleve, A. R., Simon, M. I., and Riggs, A. D., Measurements by quantitative PCR of changes in HPRT, PGK-1, PGK-2, APRT, MTase, and Zfy gene transcripts during mouse spermatogenesis, *Nucleic Acids Res.*, 18,1255, 1990.
18. Dallman, M. J., Montgomery, R. A., Larsen, C. P., Wanders, A., and Wells, A. F., Cytokine gene expression: Analysis using Northern blotting, polymerase chain reaction and *in situ* hybridization, *Immunol Rev.*, 119, 163, 1991.
19. Wiesner, R. J., Direct quantification of picomolar concentrations of mRNAs by mathematical analysis of a reverse transcription/exponential polymerase chain reaction assay, *Nucleic Acids Res.*, 20, 5863, 1992.
20. Horikoshi, T., Danenberg, K. D., Stadlbauer, T. H. W., Volkenandt, M., Shea, L. C. C., Aigner, K., Gustavsson, B., Leichman, L., Frosing, R., Ray, M., Gibson, N. W., Spears C. P., and Danenberg, P. V., Quantitation of thymidilate synthase, dihydrofolate reductase, and DT-diaphorase gene expression in human tumors using the polymerase chain reaction, *Cancer Res.*, 52, 108, 1992.
21. Noonan, K. E., Beck, C., Holzmayer, T. A., Chin, J. E., Wunder, J. S., Andrulis, I. L., Gazdar, A. F., Willman, C. L., Griffith, B., Von Hoff, D. D., and Roninson, I. B., Quantitative analysis of *MDR1* (multidrug resistance) gene expression in human tumors by polymerase chain reaction, *Proc. Natl. Acad. Sci. USA*, 87, 7160, 1990.
22. Murphy, L. D., Herzog, C. E., Rudick, J. B., Fojo, A. T., and Bates, S. E., Use of the polymerase chain reaction in the quantitation of *mdr*-1 gene expression, *Biochemistry*, 29, 10351, 1990.
23. Kinoshita, T., Imamura, J., Nagai, H., and Shimotohno, K., Quantification of gene expression by a wide range by polymerase chain reaction, *Anal. Biochem.*, 206, 231, 1992.
24. Sivitz, W. I. and Lee, E. C., Assessment of glucose transporter gene expression using the polymerase chain reaction, *Endocrinology*, 128, 2387, 1991.
25. Chelly, J., Kaplan, J. C., Maire, P., Gautron, S., and Kahn, A., Transcription of the dystrophin gene in human muscle and non-muscle tissue, *Nature*, 333, 858, 1988.
26. Wang, A. M., Doyle, M. V., and Mark, D. F., Quantitation of mRNA by polymerase chain reaction, *Proc. Natl. Acad. Sci. USA*, 86, 9717, 1989.
27. Ballagi-Pordany, A. and Funa, K., Quantitative determination of mRNA phenotypes by the polymerase chain reaction, *Anal. Biochem.*, 196, 89, 1991.
28. Becker-Andre, M. and Hahlbrock, K., Absolute mRNA quantification using the polymerase chain reaction (PCR). A novel approach by a PCR aided transcript titration assay (PATTY), *Nucleic Acids Res.*, 17, 9437, 1989.

29. Gilliland, G., Perrin, S., Blanchard, K., and Bunn, H. F., Analysis of cytokine mRNA and DNA: detection and quantitation by competitive polymerase chain reaction, *Proc. Natl. Acad. Sci. USA,* 87, 2725, 1990.
30. Mackiewicz, M., Sollars, P. J., Ogilvie, M. D., and Pack, A. I., Modulation of IL-1β gene expression in rat CNS during sleep deprivation, *NeuroReport,* 7, 529, 1996.
31. Nedelman, J., Heagerty, P., and Lawrence, F., Quantitative PCR with internal controls, *Comput. Appl. Biosci.,* 8, 65, 1992.
32. Bouaboula, M., Legoux, P., Pesseque, B., Delpech, B., Dumont, X., Piechaczyk, M., Casellas, P., and Shire, D., Standardization of mRNA titration using a polymerase chain reaction method involving co-amplification with a multispecific internal control, *J. Biol. Chem.,* 267, 21830, 1992.
33. Price, T., Aitken, J., and Simpson, E.R., Relative expression of aromatase cytochrome P450 in human fetal tissues as determined by competitive polymerase chain reaction amplification, *J. Clin. Endocrinol. Metab.*, 74, 879, 1992.
34. Uberla, K., Platzer, C., Diamanstein, T., and Blankenstein, T., Generation of competitor DNA fragments for quantitative PCR, *PCR Methods App.,* 1, 136, 1991.
35. Siebert, P. D. and Larrick, J. W., PCR MIMICs: competitive DNA fragments for use as internal standards in quantitative PCR, *Biotechniques*, 14, 244, 1993.
36. Opp, M. R., Kapas L., and Toth, L. A., Cytokine involvement in the regulation of sleep, *Proc. Soc. Exp. Biol. Med.,* 201, 16, 1992.
37. Krueger, J. M., Walter, J., Dinarello, C. A., Wolff, S. M., and Chedid, L., Sleep promoting effect of endogenous pyrogen (interleukin-1), *Am. J. Physiol.,* 246, R994, 1984.
38. Opp, M. R., Obal, F., and Krueger, J. M., Interleukin 1 alters rat sleep: temporal and dose-related effects, *Am. J. Physiol.*, 260, R52, 1991.
39. Opp, M. R. and Krueger, J. M,. Interleukin-1 receptor antagonist blocks interleukin 1-induced sleep and fever, *Am. J. Physiol.,* 260, R453, 1991.
40. Opp, M. R. and Krueger, J. M., Anti-interleukin-1β reduces sleep and sleep rebound after sleep deprivation in rats, *Am. J. Physiol.,* 266, R688, 1992.
41. Opp, M. R. and Krueger, J. M., Interleukin-1 antibodies reduce NREMS and attenuate NREMS rebound after sleep deprivation in rabbits, *Sleep Res.,* 21, 323, 1992.
42. Opp, M. R. and Krueger, J. M., Effect of an interleukin-1 receptor antagonist on recovery sleep of rabbits after total sleep deprivation, *Sleep Res.*, 20, 416, 1991.
43. Breder, C. D., Dinarello, C. A., and Saper, C. B., Interleukin-1 immunoreactive innervation of human hypothalamus, *Science,* 240, 321, 1988.
44. Farrar, W. L., Kilian, P. L., Ruff, M. R., Hill, J. M., and Pert, C. B., Visualization and characterization of interlukin 1 receptors in brain, *J. Immunol.,* 139, 459, 1987.
45. Bergmann, B. M., Kushida, C. A., Everson, C. A., Gilliland, M. A., Obermeyer, W., and Rechtschaffen A., Sleep deprivation in rats: II. Methodology, *Sleep,* 12, 5, 1989.
46. Chomczynski, P. and Sacchi, N., Single-step method of RNA isolation by guanidinium thiocyanate-phenol-chlorophorm extraction, *Anal. Biochem.,* 162, 156, 1987.
47. Minami, M., Kuraishi, Y., and Satoh, M., Effect of kainic acid on messenger RNA level of IL-1β, TNFα and LIF in the rat brain, *Biochem. Biophys. Res. Comm.,* 176, 593, 1991.
48. Bloch, W., A biochemical perspective of polymerase chain reaction, *Biochemistry,* 30, 2735, 1991.

Chapter 12

Chemical Mutagenesis and Screening for Mouse Mutations with an Altered Rest–Activity Pattern

Patrick M. Nolan, Thomas A. Houpt, and Maja Bucan

Contents

I. Introduction

The availability of model organisms harboring mutant variants of genes involved in any biological process is essential for understanding the complexity of these pro-

0-8493-3361-X/98/$0.00+$.50

cesses at a molecular and biochemical level. A dream of many neurobiologists, including sleep researchers, is to identify and isolate a candidate gene for a controlling element of a neurobiological system or a behavior, based on previous physiological, anatomical, or pharmacological studies. However, the final role and function of a particular gene can only be ascertained by the analysis of the aberrant phenotype in mice with a null mutation of this gene. The mouse is particularly useful as a model in such studies because of the potential of using targeted mutagenesis in embryonic stem cells.[1,2] Although genetic differences have been found in sleep parameters such as total and paradoxical sleep times, circadian rhythms of sleep and the diurnal ratio of total sleep,[3-6] the identification of single-gene mutations affecting sleep is hampered by the invasive techniques necessary to study sleep correlates. Nevertheless, the recent analysis of prion protein-deficient mice would support the argument that a single gene can affect specific sleep parameters.[7] Due to the paucity of candidate genes for key regulators of sleep–wakefulness, we are using random mutagenesis to screen for potential sleep mutants among lines of mice with abnormal rest:activity cycles. This approach involves mutagenesis of the whole genome with a potent chemical mutagen, such as *N*-ethyl-*N*-nitrosourea (ENU),[8,9] and the generation of a large number of progeny (Figure 12.1). Subsequent screening for the presence of behavioral anomalies may reflect the disruption of gene(s) with a general effect on daily activity cycles (for a review see reference 10).

This chapter explores current and potential applications of random chemical mutagenesis to identify novel rest:activity mutations in the mouse. We describe the mutagenesis procedures and behavioral assays and constraints of genetic analysis in the study of rodent behavior. Fully computerized and automated online systems for monitoring wheel-running behavior can be used to simultaneously screen a large number of progeny of mutagenized mice. Analysis of wheel running activity records in light:dark (LD) conditions may provide a prescreen for potential sleep–wake mutants. Based on the statistical analysis of behavioral parameters which can be extracted from the activity records of a large number of tested progeny, several potential mutants with a strikingly different behavioral profile from the norm can be identified. The heritablity of alteration in wheel running rest:activity behavior in these candidates can confirm that the behavioral anomaly is due to a genetic defect. Although monitoring wheel running activity may be sufficient to identify novel mutations affecting circadian behavior,[11,12] potential mutants will have to be examined by electroencephalographic and electromyographic recording and sleep-deprivation assays in order to test whether they affect arousal, wakefulness, or sleep (Figure 12.1). Circadian rhythms of sleep–wakefulness in general covary with wheel running rest:activity cycles. However, concomitant monitoring of EEG recordings and wheel-running has shown periods of "quiet" wakefulness during the wheel running rest phase and fragmented REM and NREM sleep during the active phase of the circadian wheel-running cycle.[13,14] In addition, exercise[15] or access to activity wheels[16,17] can modify sleep patterns in mice, indicating that a prescreen could potentially identify mutations that disrupt the coupling between locomotor activity and sleep homeostasis.

In several model organisms, such as *Drosophila melanogaster, C. elegans,* and zebrafish (references 18, 19, 20, 21 and other papers in *Development,* Vol. 123), large-

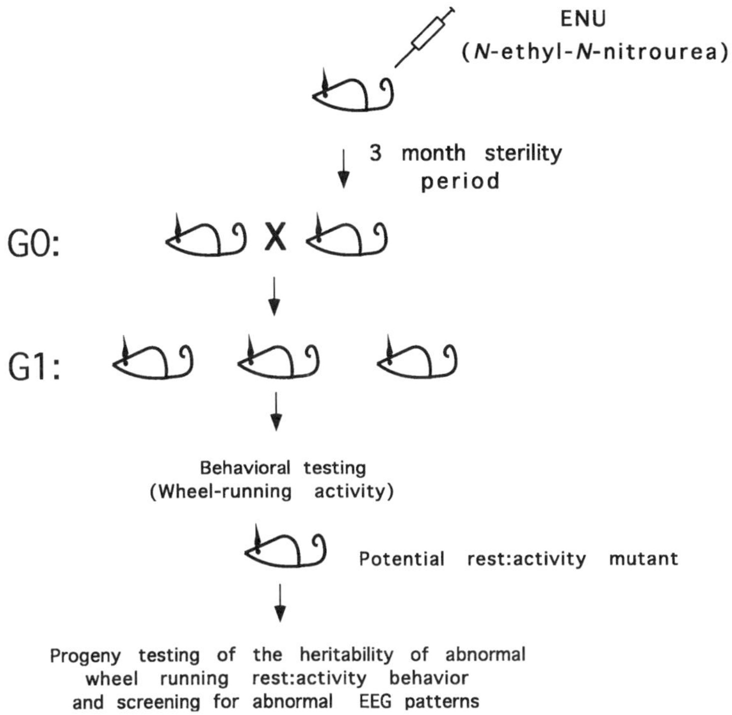

FIGURE 12.1
ENU mutagenesis. Young (8 to 10 weeks) C57BL/6J male mice are injected intraperitoneally with the chemical mutagen *N*-ethyl-*N*-nitrosourea (ENU). After a period of sterility (on average 12 to 15 weeks), mutagenized (G0) males are mated with C57BL/6J females. Progeny from this cross (G1), harboring many different induced mutations on one of their parental chromosomes, are screened for potential dominant or semidominant rest:activity mutations. Families of mice that are selected on the basis of aberrant wheel-running activity can then be tested for the heritability of the abnormal behavior and can be used to define the physiological and anatomical basis of the abnormality.

scale mutagenesis experiments for recessive and dominant mutations involve screening thousands of progeny of mutagenized individuals for the desired mutant phenotype in subsequent generations. In the mouse, extensive mutagenesis screens have not been performed due to the expense of mouse maintenance. Therefore, current efforts to recover a large number of recessive mutations are restricted to relatively short stretches of the mouse genome[22-24]; or are aimed at the identification of recessive visible mutations in a known gene.[25] However, a one-generation screen for dominant phenotypes may also provide an efficient way to recover novel mutations. For example, *Clock*, the first mutation which controls circadian behavior in the mouse,[11] and *Wheels*, a developmental mutation which alters circadian activity and responses to light (references 12, 26, and Nolan, Alavizadeh, Lo, and Bucan, unpublished), were identified in screens for mutants with a dominant mode of inheritance.

The frequency of mutations at any given gene induced by ENU is 1.5×10^{-3} per locus, per gamete[27] which, in effect, means that 1 of 750 progeny of mutagenized individuals should harbor a point mutation in any gene of interest. In the one-generation screen for dominant mutations, populations of progeny for testing are generated by mating mutagenized mice (males) with wild-type females. However, in a screen aimed at the identification of recessive mutations, where the mutant phenotype is detectable only in individuals with two copies of the mutated gene, a complex three-generation breeding scheme has to be used.

The ultimate goal of the mutagenesis screen described in this chapter is to identify single gene mutations with a profound effect on rest:activity behavior. However, in a dominant screen for behavioral mutants even mutations causing minor, but stably inherited, changes will be of great value; subtle changes in heterozygotes may be associated with a more severe phenotype in homozygotes, as shown for the *Clock* mutation.[11] To facilitate genetic analysis of these subtle mutations it is important to maintain them in a homogeneous genetic background; both mutagenized and wild-type progenitors (GO in Figure 12.1) should be of the same inbred strain. Analysis of a behavioral parameter, such as circadian period (τ_{DD}), in a large number of genetically homogeneous progeny of mutagenized mice will give a continuous distribution of τ_{DD} values, where the individual outliers may represent either 1) normal variance in the population, 2) an ENU-induced single gene mutation, or 3) may be caused by multiple mutation events in several genes. Mating of individuals with extreme behavioral phenotypes with wild-type mice will provide the critical test of heritability of the mutant phenotype, as well as its mode of inheritance

II. Methodology

A. *Mutagenesis Protocol and Breeding Scheme in a Screen for Dominant Mutations*

The procedure for chemical mutagenesis involves administration of the mutagen *N*-ethyl-*N*-nitrosourea (ENU) to 7 to 10-week-old mice with three weekly doses of 100 mg/kg by intraperitoneal injections. A repeated ENU administration protocol has proven to be highly efficient and ensures a greater survival rate for mutagenized animals.[27] Since ENU is mutagenic for mouse spermatogonial stem cells, successfully mutagenized animals will be rendered temporarily sterile. In a typical case, new rounds of spermatogenesis will cease for 10 to 15 weeks, after which surviving mutagenized spermatogonial stem cells repopulate the testis. At four weeks after administration of the mutagen, ENU-treated males are caged individually with two fertile females of the same genetic background (males remain fertile for a few weeks after ENU application). If females become pregnant during this transitional period, it will be an indication that the mutagenesis was unsuccessful and treatment will be repeated. However, restored fertility after a three-to-four month sterile period will likely indicate efficient mutagenesis. Based on the results with mutations identified to date,[28] we found that by collecting and analyzing no more than 50 to 75 generation 1 (G1) progeny from a single mutagenized male, we have a low chance of recovering the same mutation multiple times.

B. Behavioral Assays

Mice, 8 to 14 weeks old, are housed individually in cages equipped with running wheels, with food and water freely available. Cages are maintained in light-tight, ventilated chambers (10 cages to a chamber), with timed lighting. Each animal's wheel-running activity is monitored continuously, via a microswitch located on the outside of each cage. Microswitch closures for 120 cages are automatically recorded and stored on an IBM computer-based data acquisition system (developed by Ralph Mistlberger, Simon Fraser University, British Columbia, Canada). Behavioral analysis is conducted under LD (12 hours light:12 hours dark) conditions for 7 days to assess the ability of mice to synchronize or entrain to the LD cycle, followed by 3 weeks of monitoring in constant dark (DD) conditions to screen for "clock gene" mutations (i.e., mutations which either shorten or lengthen the endogenous circadian period).

C. Data Analysis of Wheel-Running Activity by the Circadia Program

A Macintosh-based interactive program, *Circadia*, is used to rapidly visualize, analyze, and quantitate the rhythmicity and activity of mutant mice. Circadian rhythms can be characterized by three quantitative parameters: period, phase, and amplitude.[29] Changes in these parameters may reveal an altered central circadian clock, altered coupling between ambient light cycles and the clock (e.g., due to changes in visual system sensitivity), or altered coupling between the clock and the physiological and behavioral processes it regulates (e.g., sleep and activity cycles). *Circadia* employs standard plotting styles and time-series algorithms that have been extensively used and validated by the circadian rhythm community.

For nocturnal mice, entrained and free-running periods should be assessed under both an ambient light-dark (LD) cycle and under constant-dark conditions (DD) (Figure 12.2 A). Wheel running activity can be visualized through ten different plotting styles with *Circadia*, from a spreadsheet view of individual data points to a strip-chart format. *Circadia* can calculate periodicity from as little as 7 to 10 consecutive days of data using several algorithms: χ^2 periodogram, which estimates period by minimizing variance when averaging data across cycles,[30] autocorrelation,[31] linear fitting to activity onsets or offsets, cosinor analysis (estimating the coefficients of a sinusoidal function by least squares fit to the entire data set).[32] These routines can be used to screen for period changes, including significant deviations of the period from 24 h under LD, significant lengthening or shortening of the period under DD, changes in ultradian or infradian rhythms, and arhythmicity.

The phase relationships of circadian rhythms are calculated with reference either to external cycles (i.e., the time of activity onset and offset relative to the time of lights-on or lights-off) or by the phase of internal rhythms relative to a reference phase marker; for example, the timing of daytime activity bouts (internal rhythms) relative to the onset of subjective night (reference phase). *Circadia*'s autocorrelation and cosinor analyses can explicitly calculate phase parameters for circadian rhythms,

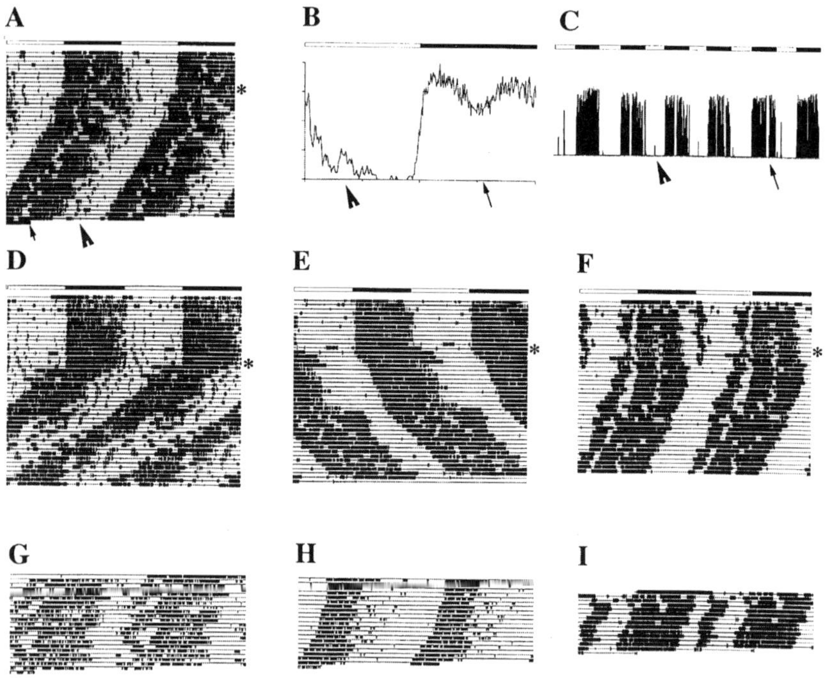

FIGURE 12.2
Wheel-running activity patterns in a typical C57BL/6J mouse and in mice with abnormal rest:activity parameters. Typical activity records for a C57BL/6J mouse A-C. Expression of wheel-running data in all three formats may be useful in the identification of potential rest:activity mutants: **A)** Double-plotted actogram raster plot of wheel-running (indicated by deflections from baseline) over 40 consecutive days under LD (days 0 to 10) and DD conditions. Dense horizontal light and dark bars represent periods of light and dark in LD, and the asterisk represents the day of transition from LD to DD. Variations in the amount, bout-distribution, periodicity, and phase of wheel-running activity are immediately obvious by visual inspection. Note the shortening of circadian period under DD, the greater precision of activity onset compared to activity offset, the brief period of inactivity in the middle of subjective night (arrow), and the short burst of activity in early subjective day (arrowhead). The ability to generate raster plots of the activity patterns of large numbers of mutants rapidly may be sufficient to identify qualitatively obvious mutants, but the statistical analysis of endogenous variation and the detection of significant changes requires additional quantitative computational analysis. **B)** The mean activity waveform (circadian time vs. mean revolutions/5 min bin) of the last 30 days under DD of the activity record in (A) averaged at the free-running period. The intersection of x and y axes represents circadian time 0 (CT0). The mean waveform is the first step in quantitating activity patterns, followed by determining period by χ^2 periodogram and rest:activity ratios by bout analysis. Arrows as in (A). **C)** The same activity record as (A) plotted in strip-chart format, as clock-time vs. revolutions/5 min bins. Although circadian parameters are less obvious, the strip chart reveals the consistent amplitude of wheel-running (~300 revolutions/5 min) during almost all bouts of activity, as well as periods of inactivity during subjective night and bouts of activity in subjective day. Arrows as in (A). **D)** Double-plotted actogram of a representative animal with a short circadian period (22.61 h) in LD and DD conditions. The asterisk represents the time of transition from LD to DD. Activity onsets are similar to (A) in LD conditions, but note the masking effect seen upon entry into DD, high levels and fragmentation of activity in subjective day, and the earlier daily onset of activity in DD relative to (A). **E)** Double-plotted actogram of a representative animal with a circadian

but are computationally intensive. Reliable phase estimates may be rapidly calculated in *Circadia* by the automatic detection of activity bout onsets; the timing of activity onsets or offsets may then be directly compared to the phase of the light cycle. Generating mean daily activity waveforms averaged and plotted at the entrained or free-running period also makes changes in circadian phase immediately obvious (Figure 12.2 B).

The amplitude of a circadian rhythm is estimated by the number of wheel-revolutions per bin at different phases of the circadian cycle, and by the total duration and amount of wheel-running activity. *Circadia* can directly quantitate wheel-running within user-specified blocks of the daily cycle. For example, the number of revolutions can be calculated for a large number of mice for each quarter of the circadian cycle, to determine if there are significant changes in late subjective night or subjective day activity, or overall rest:activity ratios. *Circadia* can perform automatic bout analysis to detect ultradian rhythms in activity timing that may be modulated by the circadian system. The initial screening may reveal alterations not in circadian rhythmicity *per se,* but in locomotor activity that would require noncircadian analyses (e.g., sleep analysis or open-field activity). Dissecting the mutational effects on locomotion and circadian rhythms is complicated by the feedback effect of activity on the circadian clock.[16]

D. Selection of Potential Rest Activity Mutants

To identify potential mutants, wheel-running activity of progeny of ENU-treated mice is examined in LD and DD conditions. Almost all G1 progeny exhibit robust activity, typical of most inbred mouse strains.[33,34] Based on the previously reported values of 23.7 ± 0.17 hours for the τ_{DD} of circadian activity in C57BL/6J mice,[11,12,34] any mouse

FIGURE 12.2 (Continued)
period greater than 24 h. Asterisk as in (D). Activity onsets are similar to (A) in LD conditions with a later daily onset of activity in DD relative to (A). No masking effect is evident and activity in subjective day is lower than in (A). Note also the periods of inactivity in subjective night. **F)** Double-plotted actogram of a representatvie animal with an advanced phase of wheel-running activity. Asterisk as in (D). Note in LD, wheel-running activity starts approximately 3 to 4 h prior to lights-off and the unusual activity onset at lights-on. The circadian period for this animal is comparable to (A) although there appear to be two activity onsets. Wheel-running activity is relatively robust, although there are brief periods of inactivity in subjective night. Subjective day activity is low. **G)** Double-plotted actogram of a representative animal with a long wheel-running activity phase (α, 18.6 h) in DD. Note that although the time between activity onsets and offsets is greater than in (A), the amplitude of wheel-running activity bouts is somewhat lower with fragmentation of wheel-running activity throughout the subjective night period. Circadian period is within the normal range for C57BL/6J mice. **H)** Double-plotted actogram of a representative animal with a short α (9.5 h) in DD. Circadian period is within the normal range for C57BL/6J mice. Activity in subjective night is robust, with several short bouts of activity in subjective day. **I)** Double-plotted actogram of a mouse in DD showing an unusual burst of activity in the subjective day period. Circadian period and activity levels in subjective night are within the normal range for C57BL/6J mice. Although it is usual to see bouts of wheel-running activity in mice at the beginning of subjective day, the activity seen here is in the middle of subjective day and is more robust than in the typical case.

with a τ_{DD} shorter than 23.10 or longer than 24.30 hours is selected for further analysis. Activity records for two mice that display significant shortening and lengthening of circadian period are illustrated in Figure 12.2 D and E. A visual observation of the onset of activity in relation to the offset of light in LD conditions allows selection of potential mutants with an early activity phase, where onset of activity precedes the offset of lights (Figure 12.2 F). Based on our data, some nongenetic factors, such as age or light intensity, may influence the precision of activity onsets.

Locomotor activity records of progeny of mutagenized mice can be analyzed to determine the variation in the length of the activity phase (α). Figure 12.2 G and H illustrate the activity records of two mice with extreme long (18 hours) or short (9 hours) activity phases (α), while circadian periods for both mice are within the normal range. Activity record Figure 12.2 I shows an individual characterized by an additional block of activity during the subjective day (under DD). Although activity at the beginning of subjective day is often noted in C57BL/6J mice, bouts of activity for this mouse occur with extreme regularity, are robust, and occur in the middle of subjective day. The circadian period for this mouse lies within the normal range.

Determining the basic circadian parameters of period, phase, and amplitude is only the beginning of characterizing a circadian system. Mutants which display hereditary alterations of greater than 2.5 SD in these parameters are subjected to more extensive, higher-resolution probing of the circadian clock. Variations in phase or levels of nocturnal and diurnal activity are behavioral variables which can be detected by monitoring robust wheel-running activity, and are relevant for the selection of candidate sleep mutants. Once potential mutants have been identified, more specialized real-time automated sleep scoring can be accomplished with available microcomputer systems.[35]

E. Initial Genetic Analysis of Selected Mutants

The next step in the characterization of selected candidates for mouse mutants with altered rest:activity behavior is a series of genetic crosses to test the following: 1) whether the mutant phenotype is transmitted to the next generation; 2) whether the mutant phenotype is due to a single gene mutation; 3) the mode of inheritance and phenotype of homozygotes; and 4) the influence of genetic background on the mutant phenotype.

The heritability of the abnormal behavior is the critical step in the characterization of potential mutants and is tested by mating the potential mutant to a wild-type mouse of the same genetic background to eliminate the influence of other genetic elements that can mask the effect of the induced mutation. In addition, each potential mutant is bred to a different inbred strain that displays similar behavioral characteristics as the original mutagenized strain. This cross will provide F1 progeny for a backcross, which will be necessary for genetic localization of the novel locus associated with the behavioral anomaly. In both crosses, the ratio of mice with altered behavior compared with the total number of progeny examined will indicate whether the inheritance follows Mendelian ratios for single-gene inheritance of a dominant trait (i.e., 50% mutant phenotype: 50% wild type based on a dominant gene). The precise ratio of mutant to normal progeny will depend on the penetrance of the mutant phenotype and/or on the relative difference between the mean value for the behavioral parameter in the particular inbred strain and that for the mutant progeny.

Although we screen progeny of mutagenized mice for a particular phenotype — altered rest:activity cycles — it is important to keep in mind that an individual sperm from a mutagenized male contains many mutational events, some of them neutral, some causing other visible phenotypes, and some that cause lethality in homozygotes. With each round of mating to a wild-type mouse (outbreeding), recombination between the mutagenized and normal genome will take place, and the desired mutant phenotype will be selected for, whereas other mutations will be eliminated. In 5 to 6 generations the genome of the mouse line selected for altered rhythms will be cleared of all but tightly linked mutations, and therefore suitable for the interbreeding that will indicate the phenotype of homozygote animals. The information concerning the phenotype in homozygotes may provide a useful insight into the nature of the genetic component that alters rest:activity behavior and other behavioral traits.

The test of the influence of genetic background on the mutant phenotype is a useful one for several reasons. First, it will indicate which mouse inbred lines are favorable for genetic mapping experiments that will be used for the chromosomal localization of the mutant locus. In addition, different expressivity/penetrance of the mutant phenotype in different genetic backgrounds can be used to identify and isolate modifiers (suppressors and enhancers) of the analyzed phenotype.

III. Discussion and Limitations of the Method

The aim of the mutagenesis screen illustrated in this chapter is to identify novel mouse mutations and corresponding disrupted genes which may serve as entry points for the construction of a molecular pathway(s) involved in governing sleep–wakefulness in mammals. The main assumption of our approach is that recordings and analyses of wheel-running rest:activity patterns allow for "behavioral high-throughput screening." The analysis of these rest:activity patterns is two-pronged: the identification of altered circadian oscillator function as revealed by circadian wheel-running parameters and the identification of altered activity parameters that may point to potential sleep mutants. We argue that further behavioral analysis, including electrophysiological recordings (by electroencephalogram and electromyogram) before and after sleep deprivation, may reveal activity mutants that harbor inherited alterations of sleep parameters.

The main limitation of a random mutagenesis screen for behavioral mutants in the mouse is the laborious nature of the large-scale screen and the capacity of the behavioral testing facility necessary to screen and analyze large numbers of mutagenized lines. This aspect restricts current efforts to one-generation screens, which will uncover only mutations with a dominant mode of inheritance. In general, dominant inheritance is rarely associated with loss-of-function mutations which, as stated before, are the most informative class of mutations in the genetic dissection of a molecular or neuronal pathway. Moreover, dominant mutations are not as frequent as recessive mutations. Therefore, only a three-generation screen and breeding to homozygosity of these randomly induced mutations will offer the opportunity to systematically isolate essential regulatory genes in a particular process such as sleep.

Any knowledge of the mode of inheritance of the phenotype of interest in other model organisms may give an important clue for the design of a mutagenesis screen/scheme in the mouse. For example, the existence of an array of semidominant alleles at the *per* locus in *Drosophila* (for a review see references 36 and 37) and the *tau* mutation in the hamster[38] provided a rationale for a dominat screen for mutations that affect circadian behavior in the mouse.[11,12] In the case of sleep anomalies, the study of narcolepsy in dogs, *canarc1*, serves as a useful example of a recessive trait which could be potentially detected in heterozygotes (G1) by pharmacological treatment; Mignot and colleagues showed that activation of cholinergic and deactivation of monoaminergic systems can induce narcoleptic/cataplectic behavior in normally asymptomatic carriers.[39,40] Neuropharmacological testing of novel behavioral mutations recovered in a random mutagenesis screen will be of great value in the initial characterization of these mutants and in the evaluation of their resemblance to corresponding human inherited disorders. Hypothetically, if a novel mouse mutant with sudden sleep attacks displays the same therapeutic response as narcoleptic subjects, this mutant will be a likely candidate for a mouse model of narcolepsy.

In a screen for a dominant mutation, the desired phenotype observed in heterozygotes (G1 progeny) may dramatically differ from the phenotype in mice with two copies of this mutant locus (homozygotes). The true nature of the induced mutation and the corresponding disrupted gene will only be ascertained following its breeding to homozygosity — intermating of two affected individuals and analysis of the mutant phenotype in their offspring. For example, we have recently shown that the *Wheels* mutation, identified in our pilot mutagenesis screen for dominant mutants with anomalies in circadian activity, is associated with embryonic lethality and anomalies in the developing forebrain in homozygotes (Nolan, Alavizadeh, Lo, and Bucan, unpublished). In the case of circadian rhythms, homozygosity at the *per*, *tim*, *tau,* and *Clock* loci results in a more severe effect on circadian rhythmicity and is not associated with lethality.[11,36-38,41] However, in the case of other dominant behavioral mutations in the mouse, it is important to await more studies to evaluate how many of them will be associated with developmental defects or lethality in homozygotes. This is particularly relevant in the case of gene(s) that may regulate such an essential biological process as sleep. The association of developmental anomalies with sleep disturbances and/or sleep apnea has been reported for several human inherited disorders, such as Smith-Magenis[42] and Treacher Collins[43] or Prader Willi syndromes.[44]

Monitoring wheel running activity in rodents provides an efficient and inexpensive screen for abnormal circadian behavior. In our effort to use the same paradigm to screen for potential sleep mutants, we are characterizing mice with abnormal values for activity phase (α) and mice with high levels of activity during the subjective day, as well as those whose rest:activity phases seem fragmented. Although, by limiting ourselves to a noninvasive prescreen we may miss subtleties in sleep–wake architecture, it is important to keep in mind that our long-term goal will be to positionally clone genes disrupted in the sleep mutations. Positional cloning involves extensive breeding and phenotypic analysis of several hundreds of mice. Therefore, at this stage of sleep genetics, sleep mutants that are "marked" by an additional phenotype and scoreable in the high throughput assay are an important advantage.

The ultimate goal of a screen for mutations with abnormal rest:activity behavior is the identification of a disrupted gene. How does one discover the molecular identity of a gene whose function is not known? The strategy that allows cloning of a gene based on its chromosomal map location is positional cloning.[45] Currently, procedures for positional cloning include genetic mapping of the locus of interest to a 0.5 cM interval (approximately 1 Mb), physical mapping, cloning of large genomic DNA fragments, and finally a search for genes of interest in the cloned region. On average, it takes two to three years to positionally clone the gene disrupted in a novel mutation in the mouse. With comprehensive physical and transcription maps, valuable resources currently being developed by the Human Genome Project, conventional positional cloning may soon be replaced by techniques that allow one to search for point mutations in candidate genes in the immediate vicinity of the mutant locus rather than search for these candidate genes in a larger chromosomal segment.[46] In other words, the availability of transcription maps and expression profiles for genes placed on these maps will allow faster identification of genes, disrupted in novel behavioral mutants in the mouse, as well as identification of their homologs in humans.

Acknowledgments

We thank Alicia Dixon for animal care, Doua Xiong and Michael Urashka for behavioral testing, Gary Pickard and Ralph Mistlberger for help and advice with the set up of the behavioral testing unit, and Dani Reed, Amita Sehgal, Larry Sanford, Sigrid Veasey, and Scott Poethig for discussions and their comments on the manuscript. These studies were supported by grants Air Force Office for Scientific Research (F49620-94-1-0234) and NIH Grant HD 28410 (to M.B.).

References

1. Capecchi, M.R., Altering the genome by homologous recombination, *Science,* 244, 1288, 1989.
2. Joyner, A.L., *Gene Targeting: A practical Approach*, Oxford University Press, New York, 1993.
3. Friedmann, J. K., A diallel analysis of the genetic underpinnings of mouse sleep, *Physiol. Behav.,* 12, 169, 1974.
4. Valatx, J. L., Bugat, R., and Jouvet, M., Genetic studies of sleep in mice, *Nature,* 238, 226, 1972.

5. Mitler, M. M., Cohen, H. B., Grattan, J., Dominic, J., Deguchi, T., Barchas, J. D., Dement, W. C., and Kessler, S., Sleep and serotonin in two strains of Mus musculus, *Pharmacol. Biochem. Behav.,* 1, 501, 1973.
6. Oliverio, A. and Malorni, W., Wheel running and sleep in two strains of mice: plasticity and rigidity in the expression of circadian rhythmicity, *Brain Res.,* 163, 121, 1979.
7. Tobler, I., Deboer, T., and Fischer, M., Sleep and sleep regulation in normal and prion protein-deficient mice, *J. Neurosci.* 17, 1869, 1997.
8. Russell, W.L., Kelly, E.M., Hunsicker, P.R., Bangham, J.W., Maddux, S.C., and Phipps, E.L., Specific-locus test shows ethylnitrosurea to be the most potent mutagen in the mouse, *Proc. Natl. Acad. Sci. USA,* 76, 5818, 1979.
9. Rinchik, E.M., Chemical mutagenesis and fine-structure functional analysis of the mouse genome, *Trends Genet.,* 7, 15, 1991.
10. Takahashi, J.S., Pinto, L.H., and Vitaterna, M.H., Forward and reverse genetic approaches to behavior in the mouse, *Science,* 264, 1724, 1994.
11. Vitaterna, M.H., King, D.P., Chang, A-M., Kornhauser, J.M., Lowrey, P.L., McDonald, J.D., Dove, W.F., Pinto, L.H., Turek, F.W., Takahashi, J.S., Mutagenesis and Mapping of a Mouse Gene, *Clock*, Essential for Circadian Behavior, *Science,* 264, 719, 1994.
12. Pickard, G. E., Sollars P. J., Rinchik, E. M., Nolan, P. M., and Bucan, M., Mutagenesis and behavioral screening for altered circadian activity identifies the mouse mutant, *Wheels (Whl), Brain Research,* 705, 255, 1995.
13. Richardson, G. S., Moore-Ede, M. C., Czeisler, C. A., and Dement, W. C., Circadian rhythms of sleep and wakefulness in mice: analysis using long-term automated recording of sleep, *Am. J. Physiol.* 248, R320, 1985.
14. Mitler, M. M., Lund, R., Sokolove, P. G., and Pittendrigh, C. S., Sleep and activity rhythms in mice: a description of circadian patterns and unexpected disruptions in sleep, *Brain Res.,* 131, 129, 1977.
15. Mistlberger, R. A., Bergman, B. M., and Rechtschaffen, A., Period-amplitude analysis of rat electroencephalogram: effects of sleep deprivation and exercise, *Sleep,* 10, 508, 1987.
16. Edgar, D. M., Martin, C. E., and Dement, W. C., Activity feedback to the mammalian circadian pacemaker: influence on observed measures of rhythm period length, *J. Biol. Rhythms,* 6, 185, 1991.
17. Welsh, D. K., Richardson, G. S., and Dement, W. C., Effect of running wheel availability on circadian patterns of sleep and wakefulness in mice, *Physiol. Behav.,* 43, 771, 1988.
18. Nüsslein-Volhard, C. and Wieschaus, E., Mutations affecting segment number and polarity in *Drosophila., Nature,* 287, 795, 1980.
19. Hirsh, D. and Vanderslice, R., Temperature-sensitive developmental mutants of *caenorhabditis elegans, Dev. Biol.,* 49, 220, 1976.
20. Haffter, P., Granato, M., Brand, M., Mullins, M. C., Hammerschmidt, M., Kane, D. A., Odenthal, J., van Eeden, F. J. M., Jiang, Y.-J., Heisenberg, C.-P., Kelsh, R. N., Furutani-Seiki, M., Vogelsang, E., Beuchle, D., Schach, U., Fabian, C., and Nusslein-Volhard, C., The identification of genes with unique and essential functions in the development of the zebrafish, *Danio rerio, Development,* 123, 1, 1996.

21. Driever, W., Solnica-Krezel, L., Schier, A. F., Neuhauss, S. C. F., Malicki, J., Stemple, D. L., Stainier, D. Y. R., Zwartkruis, F., Abdelilah, S., Rangini, Z., Belak, J., and Boggs, C., A genetic screen for mutations affecting embryogenesis in zebrafish, *Development*, 123, 37, 1996.
22. Bode, V.C., Ethylnitrosourea mutagenesis and the isolation of mutant alleles for specific genes located in the *T* region of mouse chromosome 17, *Genetics,* 108, 457, 1984.
23. Shedlovsky, A., Guenet, J-L., Johnson, L.L., and Dove, W.F., Induction of recessive lethal mutations in the T/*t-H-2* region of the mouse genome by a point mutagen, *Genet. Res.*, 47, 135, 1986.
24. Rinchik, E.M., Carpenter, D.A., and Selby, P.B., A strategy for fine-structure functional analysis of a 6- to 11-centimorgan region of mouse chromosome 7 by high-efficiency mutagenesis, *Proc. Natl. Acad. Sci. USA*, 87, 896, 1990.
25. Cordes, S.P. and Barsh, G.S., The Mouse segmentation gene *kr* encodes a novel basic domain-leucine zipper transcription factor, *Cell,* 79, 1025, 1994.
26. Nolan, P. M., Sollars, P. J., Bohne, B. A., Ewens, W. J., Pickard, G. E., and Bucan, M., Heterozygosity mapping of partially congenic lines: mapping of a semidominant neurological-mutation, *Wheels (Whl),* on mouse chromosome 4, *Genetics,* 140, 245, 1995.
27. Hitotsumachi, S., Carpenter, D.A., and Russell, W.L., Dose repetition increases the mutagenic effectiveness of *N*-ethyl-*N*-nitrosurea in mouse spermatogonia, *Proc. Natl. Acad. Sci. USA,* 82, 6619, 1985.
28. Nolan P.M., Kapfhamer D., and Bucan, M., Procedures for the identification of dominant behavioral mutations in mice using ENU mutagenesis, *Methods: A Companion to Methods in Enzymology. Functional Genomics in the Mouse*, submitted.
29. Moore-Ede, M.C., Sulzman, F.M., and Fuller, C.A., *The Clock That Times Us*, Harvard University Press, Cambridge, 1982, Chaps. 1 and 2.
30. Chatfield, C., *The Analysis of Time Series: An Introduction*, New York, Chapman and Hall, 1984.
31. Bingham, C., Arbogast, B., Guillaume, G.C., Lee, J.K., and Halberg, F., Inferential statistical methods for estimating and comparing cosinor parameters, *Chronobiologia,* 9, 397, 1982.
32. Dorrscheift, G.J. and Beck, J., Advanced methods for evaluating chracteristic parameters (tau, alpha, ro) of circadian rhythms, *J. Math. Biol.,* 2, 107, 1975.
33. Pittendrigh, C.S. and Daan, S., A functional analysis of circadian pacemakers in nocturnal rodents. I. The stability and lability of spontaneous frequency, *J. Comp. Physiol.,* 106, 223, 1976.
34. Schwartz, W.J. and Zimmerman, P., Circadian timekeeping in BALB/c and C57BL/6 inbred mouse strains, *J. Neurosci.,* 10, 3685, 1990.
35. Van Gelder, R.N., Edgar, D.M., and Dement, W.C., Real-time automated sleep scoring, validation of a microcomputer-based system for mice, *Sleep*, 14, 48, 1991.
36. Dunlap, J.C., Genetic analysis of circadian rhythms, *Ann. Rev. Physiol.*, 55, 683, 1993.
37. Hall, J.C., Genetics of circadian rhythms, *Ann. Rev. Genet.,* 24, 659, 1993.
38. Ralph, M.R. and Menaker, M., A mutation of the circadian system in golden hamsters, *Science,* 241, 1225, 1988.

39. Lucas, E.A., Foutz, A.S., Dement W.C., and Mitler, M.M., Sleep cycle organization in narcoleptic and normal dogs, *Physiol. Behav.,* 23, 737, 1979.
40. Mignot E., Nishino, S., Sharp, L.H., Arrigoni, J., Siegal, J.M., Reid, M.S., Edgar, D.M., Ciaranello, R.D., and Dement, W.C., Heterozygosity at the *canarc-1* locus can confer susceptibility for narcolepsy: Induction of cataplexy in heterozygous asymptomatic dogs after administration of a combination of drugs acting on monoaminergic and cholineric systems, *J. Neurosci.,* 13, 1057, 1993.
41. Sehgal, A., Price, J., and Young, M.W., Circadian behavioral rhythms and molecular oscillations of *per* RNA abolished by a new *Drosophila* mutation, *timeless*, *Science,* 263, 1603, 1994.
42. Greenberg, F., Lewis, R.A., Potocki, L., Glaze, D., Parke, J., Killian, J., Murphy, M.A., Williamson, D., Brown, F., Dutton, R., McCluggage, C., Friedman, E., Sulek, M., and Lupski, J.R., Multi-Disciplinary clinical study of Smith-Magenis Syndrome (Deletion 17p11.2), *Am. J. Med. Genet.,* 62, 247, 1996.
43. Colmenero, C., Esteban, R., Albarino, A.R., and Colmenero, B., Sleep Apnoea Syndrome associated with maxillofacial abnormalities, *J. Laryngol. Otol.,* 105, 94, 1991.
44. Clarke, D.J., Waters, J., and Corbett, J.A., Adults with Prader-Willi Syndrome: abnormalities of sleep and behaviour, *J. Roy. Soc. Med.*, 82, 21, 1989.
45. Collins, F.S., Positional cloning: Let's not call it reverse anymore, *Nature Genetics,* 1, 3, 1992.
46. Schuler, G.D., Boguski, M.S., Stewart, E.A., Stein, L.D., Gyapay, G., Rice, K., White, R.E., Rodriguez-Tome, P., Aggarwal, A., Bajorek, E., Bentolila, S., Birren, B.B., Butler, A., Castle, A.B., Chiannilkulchai, N., Chu, A., Clee, C., Cowles, S., Day, P.J., Dibling, T., Drouot, N., Dunham, I., Duprat, S., East, C., Edwards, C., Fan, J.-B., Fang, N., Fizames, C., Garrett, C., Green, L., Hadley, L., Harris, M., Harrison, P., Brady, S., Hicks, A., Holloway, E., Hui, L., Hussain, S., Louis-Dit-Sully, C., Ma, J., MacGilvery, A., Mader, C., Maratukulam, A., Matise, T.C., McKusick, K.B., Morissette, J., Mungall, A., Muselet, D., Nusbaum, H.C., Page, D.C., Peck, A., Perkins, S., Piercy, M., Qin, F., Quackenbush, J., Ranby, S., Reif, T., Rozen, S., Sanders, C., She, X., Silva, J., Slonim, D.K., Soderlund, C., Sun, W.-L., Tabar, P., Thangarajah, T., Vega-Czarny, N., Vollrath, D., Voyticky, S., Wilmer, T., Wu, X., Adams, M.D., Auffray, C., Walter, N.A.R., Brandon, R., Dehejia, A., Goodfellow, P.N., Houlgatte, R., Hudson, Jr., J. R., Ide, S.E., Iorio, K.R., Lee, W.Y., Seki, N., Nagase, T., Ishikawa, K., Nomura, N., Phillips, C., Polymeropoulos, M.H., Sandusky, M., Schmitt, K., Berry, R., Swanson, K., Torres, R., Venter, J.C., Sikela, J.M., Beckmann, J.S., Weissenbach, J., Myers, R.M., Cox, D.R., James, M.R., Bentley, D., Deloukas, P., Lander, E.S., and Hudson, T.J., A gene map of the human genome, *Science*, 274, 540, 1996.

Chapter 13

Reverse Transcription mRNA Differential Display: A Systematic Molecular Approach to Identify Changes in Gene Expression Across the Sleep–Waking Cycle

Maria Pompeiano, Chiara Cirelli, and Giulio Tononi

Contents

0-8493-3361-X/98/$0.00+$.50

I. Introduction

The analysis of gene expression has become an important tool for understanding physiological and pathological processes, including development, plasticity, response to injury, carcinogenesis, regeneration, and death. Recently, it has been demonstrated that widespread changes in gene transcription occur during the physiological sleep–waking cycle and after sleep deprivation.[3,12]

Three approaches commonly used to systematically investigate changes in gene expression are plus–minus hybridization, subtractive hybridization, and reverse transcription mRNA differential display. The classical plus–minus hybridization is straightforward and allows easy preparation of complex cDNA probes from two sources of neural tissue (for example, A and B) by performing cDNA synthesis in the presence of α^{32}P-dATP. Filter lifts of a cDNA library from tissue A are hybridized, in duplicate, with either tissue A or tissue B probes. The signal intensity of each clone detected on a filter reflects the relative abundance of each mRNA. This method, however, gives high background and only detects relatively abundant mRNAs (>0.02 to 0.06% total mRNA mass). This is a serious problem because most brain-specific mRNAs are thought to be expressed at very low levels (1 to 20 copies per cell; on average, 0.02% total mRNA mass). These rare mRNAs are more difficult to isolate, but they are also more likely to be differentially expressed.

Subtractive hybridization (SH) can isolate relatively rare differentially expressed mRNAs (0.001% total mRNA mass). Radiolabeled single-stranded cDNAs from tissue A are hybridized in solution with an excess amount of driver poly(A)+ RNA prepared from tissue B. Column chromatography is used to isolate the single-stranded cDNA fraction. Hybridization of the subtracted probe to filters from a tissue A cDNA library should only detect tissue A-specific clones. This method is technically demanding because it requires isolation of poly(A)+ RNA and construction of cDNA libraries. A disadvantage is that the amount of tissue required is very high, so that pooling of RNA or cDNA from different animals is often necessary. Pooling makes it impossible to compare intra- versus intergroup variability. SH is a kinetic reaction that never reaches completion and, therefore, not all cDNAs shared between two tissues are subtracted out. Furthermore, some genes may be lost during the construction of the cDNA library. Other problems with SH are that it is qualitative and that multiple comparisons are difficult and time-consuming because the uniquely expressed mRNAs are identified only in one selected tissue at a time.

Reverse transcription mRNA differential display (mRNA-DD) is a polymerase chain reaction (PCR)-based technique that allows a near-exhaustive comparison of gene expression from small samples of tissue obtained from multiple experimental groups. The essence of mRNA-DD is to use an anchored oligo-dT primer which anneals to the beginning of a subpopulation of the poly(A) tails of mRNAs for reverse transcription. These 3′-anchored primers are used in conjunction with arbitrarily defined sequences (5′ arbitrary primers) for the subsequent PCR amplification. Amplified, radiolabeled cDNA fragments of 3′ termini, up to 1 to 1.2 kb in size, are then separated on a denaturing polyacrylamide gel. The use of an appropriate number of primer combinations allows a side-by-side comparison between multiple experimental groups of nearly all the mRNAs (5 to 15,000) of a given tissue. The general pattern of bands has been shown to be reproducible. Since 1992, the original method[8] has been modified with respect to the primers,[1,6,9,10,19] radioactive deoxyribonucleoside triphosphates,[11,15,16] purification and cloning of the amplification products,[2,19] and direct sequencing of the bands.[14,18] Recently, a nonradioactive method has also been developed to visualize the cDNA fragments obtained from mRNA-DD.[4] The type of primers that are used is one of the crucial variables of the technique. Originally,[8] 3′ primers were two-base anchored oligo-dT primers $T_{12}MN$, i.e., 12 T's with two additional 3′ bases which provided specificity (where M can be A, C, or G, and N can be A, T, C, or G). In this case 12 different 3′ primers were used. Subsequently[9] it was shown that the penultimate base M can exhibit considerable degeneracy, while it is the last base N that provides most of the specificity. Therefore, only four two-base anchored degenerate primers that differed only in the last 3′ base were used. Recently,[10] three one-base anchored primers (H-T_{11}G, H-T_{11}A, HT_{11}C; where H is a restriction enzyme site) have been successfully employed. This new optimization reduces the number of reverse transcription reactions needed for each RNA sample, and it also minimizes the under-representation of certain RNA species due to the degeneracy of the primers.

We used a modified version of mRNA-DD to look at potential differences in gene expression between sleep and waking.[13] Three groups of rats were used. The first group of animals (sleep, S) was composed of rats that had been spontaneously asleep for three hours during the light period. The second group (sleep-deprived, SD) had been kept awake during the same period by gentle handling every time the electroencephalogram indicated a beginning of synchronized activity. The third group (waking, W) had been spontaneously awake for three hours during the dark period. These three groups were chosen in order to restrict the search for differentially expressed genes to those that are more directly related to sleep and waking *per se*, as opposed to those that may be related to circadian factors or handling. As discussed in Chapter 5 (Cirelli et al., this book), by using *in situ* hybridization and immunocytochemistry, we found a significant increase of both mRNA and protein levels of the immediate early genes c-fos and NGFI-A in several brain regions during the waking state with respect to the sleeping state.[3,12] To test the sensitivity and reproducibility of differential display in detecting changes in mRNA levels across the sleep–waking cycle, we designed specific 5′ primers for c-fos and NGFI-A. As shown in Figure 13.1, we found that mRNA-DD is able to detect changes in the expression of these IEGs similar to those that we previously detected with radioactive *in situ* hybridization. Subsequently, several other combinations of 3′-anchored primers and 5′ arbitrary primers have been employed to identify potential genes that are differentially regulated across the sleep–waking cycle.[13]

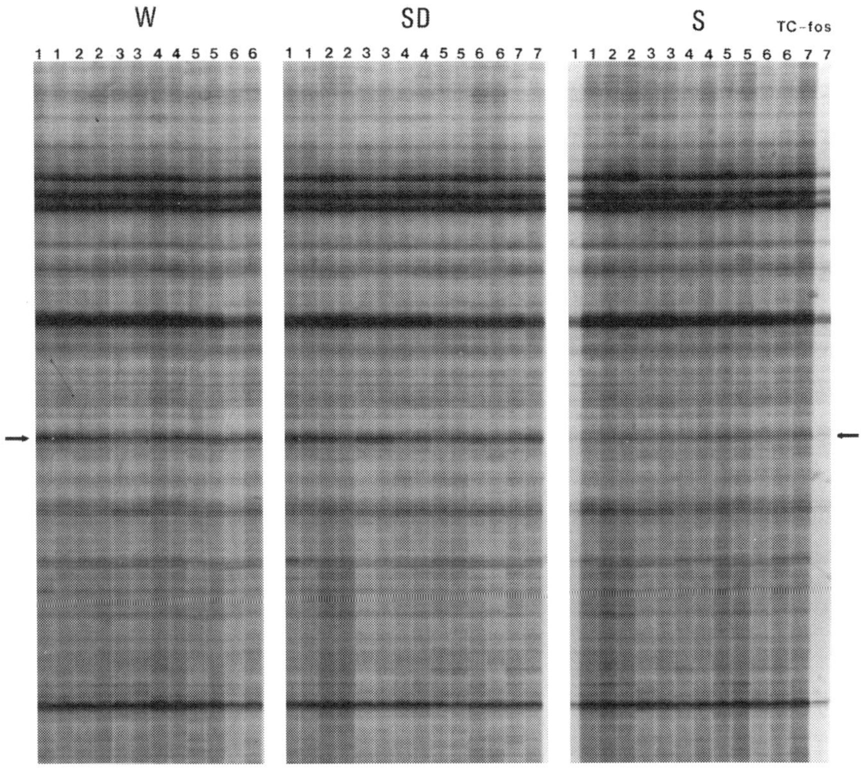

FIGURE 13.1
Total RNA from the left cerebral cortex of each animal (7 S, 7 SD, 6 W rats) was reverse transcribed with the 3′ anchored primer E1T_{12}C. DD–PCR was then performed using the same 3′ primer and a 5′ primer specifically designed for c-fos (CTG ACT CAC TGA GCT CGC CC). For each rat, two PCR reactions were performed from the same reverse transcription reaction. The products of each PCR reaction (two for each rat) were separated on a 6% acrylamide/urea gel (Genomyx). A portion of the gel, centered on the band corresponding to c-fos (arrows) is shown. Each lane on the gel represents one PCR reaction from one rat. The band indicated by the arrows was cut, reamplified, and subcloned. Sequence analysis showed that it was a 300 bp cDNA complementary to the rat c-fos mRNA. Semiquantitative analysis from a phosphorimager scan showed that the signal intensity of the band was higher in W and SD rats with respect to S rats, and the difference was statistically significant (Kruskal-Wallis test).

II. Protocol

A. Primers

Primers are synthesized on a DNA synthesizer (Beckman, Fullerton, CA) and purified by using a cleavage and deprotection kit (Beckman) according to the manufacturer's instructions. The following primers are used: 3′-anchored primers: E1-T_{12}M, where E1 = CGGAATTCGG and M= C, G, or A; 5′ arbitrary primers: E2-AP, where E2 = CGTGAATTCG, and AP is a sequence of ten base pairs (see reference 1); 5′ primer specific for c-fos: CTG ACT CAC TGA GCT CGC CC; 5′

primer specific for NGFI-A: TAG GTC AGA TGG AAG ATC TC. The 3′-anchored primer for both c-fos and NGFI-A is E1-T_{12}C. The addition of ten base pairs encoding a restriction enzyme site (E1 and E2) to both the anchored and arbitrary primers has been shown to increase reproducibility, reduce false positives, and aid cloning.[10,19] In our experience, we found no differences in the number of bands obtained with specific primer pairs (e.g., TC-fos) as compared to arbitrary primers (e.g., TC-arbitrary primer n. 5). Importantly, the pattern of bands obtained with a given primer combination is reproducible with different RNA samples and with the same RNA sample in separate experiments.

B. *Reverse Transcription Reaction*

RNA is extracted from the left cerebral cortex using RNA-zol (Biotecx, Houston, TX) according to the manufacturer's instructions. Two μg RNA (1 μg/μl) are used in each reverse transcription reaction.

1. Prepare reaction mix 1 (quantities given are enough for 10 tubes): 92 μl DEPC-treated water, 20 μl anchored primer (10 μ*M*). Mix well.
2. Aliquot 11.2 μl of mix 1 into each RT reaction tube that contains 2 μg of RNA and mix well by gentle vortexing. Place the tubes at 70°C for 10 min, then keep them on ice until required for the next step.
3. Prepare reaction mix 2 (quantities given are for 10 tubes): 40 μl 5X first-strand reaction buffer (Gibco BRL, Gaithersburg, MD), 2 μl DTT (100 μ*M*), 16 μl dNTP mix (250 μ*M*; PCR nucleotide mix, 10 m*M* each dNTP, Boehringer, Indianapolis, IN).
4. Add 5.8 μl of reaction mix 2 to each RT tube on ice. Place the tubes at 37°C for 10 min. Then add 1 μl of reverse transcriptase (M-MLV Reverse Transcriptase, Gibco BRL) to each tube. Continue the incubation at 37°C for another 60 min. Store the tubes at –80°C until required for differential display.
5. Before using for differential display, the cDNA resulting from these reactions should be diluted 1:10 in DEPC-treated water. These diluted reactions are then stored at –20°C.

C. *PCR*

1. The RT reactions (performed on 2 μg RNA from the tissue) should be diluted 1:10 in DEPC-treated water before use in these PCR reactions.
2. On ice, prepare enough reaction mix for the number of samples to be analyzed. We perform two PCR reactions for each animal. One control tube contains the PCR mix without any RT product. The following PCR mix is sufficient for 10 reactions: 95 μl DEPC-treated water, 20 μl 10X enzyme buffer, 20 μl arbitrary primer (4 μ*M*), 20 μl anchored primer (4 μ*M*), 18.4 μl dNTP (25 μ*M*), 4 μl PCR enzyme (Expand High Fidelity, Boehringer), 2.5 μl of [(α^{33}P]-dATP (1000 to 3000Ci/mmol, 10 mCi/ ml, Du Pont, Boston, MA). ^{33}P is preferable over both ^{32}P and ^{35}S for safety reasons (see references 11 and 16).
3. To each of the PCR reaction tubes, add 18 μl of the PCR mix. Then add 2 μl of each of the diluted RT reactions to the appropriate PCR tube.

4. After each of these additions is complete, the tubes should be vortexed at medium speed and placed back on ice. Collect all the reagents at the bottom of the tube by briefly centrifuging the tubes at 3500 rpm at 4°C.
5. Add 25 µl of mineral oil to each tube. PCR conditions: 4 cycles of 94°C for 1 min, 45°C for 5 min, 72°C for 90 sec, and then 30 cycles of 94°C for 30 sec, 58°C for 2 min, 72°C for 1 min. The final extension cycle is 72°C for 5 min. The samples are stored at –80°C until needed.

PCR controls include both the omission of reverse transcriptase in the RT step, and of cDNA template (RT mixture) in the PCR step.

D. Running the mRNA-DD Gel — Identifying Bands of Interest

The sequencing system used in our laboratory has been specifically designed for differential display (Genomix, Foster City, CA). It allows to detect up to 150 bands on each gel, ranging from 0.2 to 1.2 kb in size.

1. Before running the samples, add to each tube 6 µl of DNA loading buffer (950 µl formamide, 20 µl 0.5 *M* EDTA, 20 µl 4% bromophenol blue, 20 µl 4% xylene cyanol).
2. Heat the samples for 10 to 15 min at 85°C and then immediately place them on ice.
3. Load 5 µl of each sample. Initially, the samples are separated on a 6% acrylamide/urea gel (Genomyx; Figure 13.1). To increase the separation of the bands so that the cDNA of interest can be cut from the gel, a 4.5% acrylamide/ urea gel (Genomyx) is then used.
4. Dried gels are exposed overnight to X-ray film and subsequently scanned using a phosphorimager (Molecular Dynamics, Sunnyvale, CA).
5. Bands that are candidates for differential expression between different groups, or conditions can be identified by visually examining the X-ray film. We recommend a semiquantitative evaluation that can be obtained from the phosphorimager scan. For each band of interest, the average volume of that band can be determined for each lane and the background subtracted. A nonparametric, one-way analysis of variance (e.g., Kruskal-Wallis test) can be performed to determine whether the differences among groups or conditions are statistically significant.

E. Reamplification, Cloning, Sequencing — Confirmation of the Bands

The bands of interest can be cut from the gel, reamplified by PCR using the same primers used for the differential display PCR, subcloned, and then sequenced according to standard procedures. Depending on the abundance of the transcript and on the relative change in its expression, Northern blot analysis or ribonuclease protection assays can be used to confirm the mRNA-DD results. Radioactive *in situ* hybridization can also be used to confirm the difference in mRNA levels and, in the case of unknown transcripts, to describe their anatomical distribution in the brain.

III. Discussion

The technique of mRNA-DD has been successfully used to study differential gene expression in several experimental protocols, such as cardiac rejection,[17] cocaine administration,[5] and nerve injury.[7] In our opinion, mRNA-DD has several advantages over SH: 1) mRNA-DD requires as little as 0.02 μg of total mRNA so that pooling is not necessary even for small brain regions; 2) it is possible to perform multiple comparisons simultaneously; 3) the comparison between intragroup and intergroup variability allows a rational selection of bands of interest for subsequent cloning (in our case, bands related to sleep and waking and not related to circadian and/or handling factors); 4) mRNA-DD can be semiquantitative; and 5) it is relatively straightforward to obtain the pattern of bands. However, it must be kept in mind that mRNA-DD requires many primer combinations to adequately represent the 5,000 to 15,000 mRNAs expressed in a given tissue (approximately 75 primer sets are required, three one-base anchored primers combined with 20 to 25 arbitrary primers; see reference 1).

An important question is whether this technique can detect rare mRNAs. For an mRNA to be detected it must be prevalent enough to produce a signal in the autoradiograph and contain a sequence in its 3′ 1000 to 1200 nucleotides that can serve as a site for mismatched primer binding and priming. Liang et al.[9] showed in at least one case that a rare mRNA could be detected by using a 5′ primer that matched in all ten bases. However, it is still unclear whether rare mRNAs can be detected using 5′ primers having two or more mismatches. On the other hand, the more prevalent an mRNA species is, the more likely it is to generate a product. Abundant mRNAs may thus give bands with many different arbitrary primers (redundancy). This problem seems to have been reduced through the use of 3′ one-base anchored primers. In addition, when it is crucial to isolate moderate- to low-abundance mRNAs and to avoid the isolation of high abundance transcripts (such as ribosomal and mitochondrial sequences), it is helpful to use a higher initial PCR annealing temperature, (e.g., 50°C), to use primers with a relatively low GC content (<50%), and to select for further analysis those bands that are dependent on both (arbitrary and anchored) primers for amplification.[6] A potential problem when using the original protocol by Liang et al.[9] is that there is not a true linearity in the PCR-based amplification of mRNAs. This problem is less worrisome when using longer primers and higher stringency PCR conditions. Another problem that has been reported is that more than one PCR product can sometimes be generated during reamplification of a band. Using our conditions, we have almost never encountered this problem, probably because of the very high resolution and excellent separation of the bands that we obtain on our gels. Finally, in early experiments using mRNA-DD, a significant problem was the high number of false positives, i.e., differentially expressed bands that could not be confirmed by Northern blot analysis. We substantially reduced this problem by performing, for each combination of primers, duplicate PCR reactions for each subject (7 sleep, 7 sleep-deprived, and 6 spontaneously awake rats) and by only selecting bands for further analysis that are consistently present (or absent) in all the subjects within a group. In addition, ribonuclease protection assay and radioactive *in situ* hybridization are clearly preferable for confirming the bands because of their high sensitivity.

References

1. Bauer, D., Müller, H., Reich, J., Riedel, H., Ahrenkiel, V., Warthoe, P., and Strauss, M., Identification of differentially expressed mRNA species by an improved display technique (DDRT-PCR), *Nucl. Acid Res.*, 21, 4272, 1993.
2. Callard, D., Lescure, B., and Mazzolini, L., A method for the elimination of false positives generated by the mRNA differential display technique, *BioTechniques,* 16, 1096, 1993.
3. Cirelli, C., Pompeiano, M., Tononi, G., Sleep deprivation and c-fos expression in the rat brain, *J. Sleep Res.*, 4, 92, 1995.
4. Doss, R.P., Differential display without radioactivity — A modified procedure, *BioTechniques,* 21, 408, 1996.
5. Douglass, J., McKinzie, A.A., and Couceyro, P., PCR differential display identifies a rat brain mRNA that is transcriptionally regulated by cocaine and amphetamine, *J. Neurosci.*, 15, 2471, 1995.
6. Ikonomov, O.C. and Jacob, M.H., Differential display protocol with selected primers that preferentially isolates mRNAs of moderate- to low-abundance in a microscopic system, *BioTechniques,* 20, 1030, 1996.
7. Kiryu, S., Yao, G. L., Morita, N., Kato, H., and Kiyama, H., Nerve injury enhances rat neuronal glutamate transporter expression: identification by differential display PCR, *J. Neurosci.*, 15, 7872, 1995.
8. Liang, P. and Pardee, A. B., Differential display of eukaryotic messenger RNA by means of the polymerase chain reaction, *Science,* 257, 967, 1992.
9. Liang, P., Averboukh, L., and Pardee, A. B., Distribution and cloning of eukaryotic mRNAs by means of differential display: refinements and optimization, *Nucl. Acid Res.*, 21, 3269, 1993.
10. Liang, P., Zhu, W., Zhang, X., Guo, Z., O'Connell, P., Averboukh, L., Wang, F., and Pardee, A. B., Differential display using one-base anchored oligo-dT primers, *Nucl. Acid Res.*, 22, 5763, 1994.
11. Liang, P. and Pardee, A. B., Response to Trentmann et al., *Science* 267, 1187, 1995.
12. Pompeiano, M., Cirelli, C., and Tononi, G., Immediate-early genes in spontaneous wakefulness and sleep: Expression of c-fos and NGFI-A mRNA and protein, *J. Sleep Res.*, 3, 80, 1994.
13. Pompeiano, M., Cirelli, C., and Tononi, G., Changes in gene expression between wakefulness and sleep revealed by mRNA differential display, *Soc. Neurosci. Abstr.*, 273.3, 1996.
14. Reeves, S.A., Rubio, M.-P., and Louis, D.N., General method for PCR amplification and direct sequencing of mRNA differential display products, *BioTechniques,* 18, 18, 1995.

15. Tokuyama, Y. and Takeda, J., Use of 33P-labeled primer increases sensitivity and specificity of mRNA differential display, *BioTechniques,* 18, 424, 1995.
16. Trentmann, S. M., van der Knaap, E., and Kende, H., Alternatives to 35S as a label for the differential display of eukaryotic messenger RNA, *Science*, 267, 1186, 1995.
17. Utans, U., Liang P., Wyner, L.R., Karnovsky, M.J., and Russell, M.E., Chronic cardiac rejection: identification of five upregulated genes in transplanted hearts by differential mRNA display, *Proc. Natl. Acad. Sci. USA,* 91, 6463, 1994.
18. Wang, X. and Feuerstein, G.Z., Direct sequencing of DNA isolated from mRNA differential display, *BioTechniques,* 18, 448, 1995.
19. Zhao, S., Ooi, S. L., and Pardee, A. B., New primers improves precision of differential display, *Biotechniques*, 18, 842, 1995.

Chapter 14

In Situ Hybridization of Messenger RNA in Sleep Research

Tarja Porkka-Heiskanen, Jussi Toppila, and Dag Stenberg

Contents

0-8493-3361-X/98/$0.00+$.50

I. Introduction

In situ hybridization is one of the methods by which messenger RNA can be detected. A specific mRNA is detected by allowing a probe, i.e., a short, complementary DNA or RNA strand, to bind to the mRNA. For visualization the probe is labeled by an enzyme or by a radioactive isotope. *In situ* means that the labeled probe is introduced over a very thin (5 to 20 μm) tissue section and allowed to bind to the mRNA molecules that are in the cells of the section. When the cells are visualized, one gains information about the precise location of cells that contain this mRNA, in addition to an estimate of the amount of the mRNA.

We have used *in situ* hybridization to detect changes in amounts of mRNA of some genes that might be affected by sleep deprivation (both selective REM sleep deprivation and total sleep deprivation). For us, the sensitivity and specificity of this method were its main attractions. We have also wanted to document the fact that sleep loss — often regarded merely as a minor irritant — has detectable effects on cellular function at the most basic level: gene expression.

We measured growth hormone-releasing hormone (GHRH), somatostatin, and galanin mRNAs in different hypothalamic nuclei. For simultaneous quantification and localization of the mRNA molecules, *in situ* hybridization was the optimal method. Both total sleep deprivation and selective REM sleep deprivation increased the amount of somatostatin mRNA in the arcuate nucleus of rat hypothalamus.[1,2] Somatostatin mRNA also increased in the periventricular nucleus after REM sleep deprivation (Figure 14.1). GHRH mRNA increased during total sleep deprivation but decreased as a consequence of specific REM sleep deprivation in the paraventricular nucleus. In the arcuate nucleus, sleep deprivation did not affect the amount

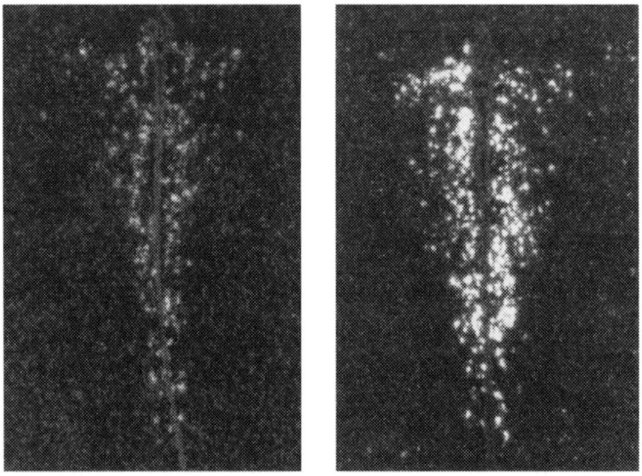

FIGURE 14.1
Effect of REM sleep deprivation on somatostatin mRNA in periventricular nucleus of rat hypothalamus. Visualization in dark field microscopy after *in situ* hybridization with a 48mer ^{35}S oligonucleotide probe for rat somatostatin mRNA. Left: control rats. Right: after 72 h of REM sleep deprivation on small platforms.

of GHRH mRNA.[1,2] Galanin mRNA increased during REM sleep deprivation in the periventricular nucleus and medial preoptic area,[3] but total sleep deprivation did not affect the amount of galanin mRNA in these same areas, indicating an effect specific to REM sleep loss or to the platform deprivation procedure used.[4] Sleep deprivation thus affects differentially the gene expression of these putative hypnogenic peptides, and this is presumably followed by corresponding changes in peptide synthesis and in physiological effects.

We have also found that the mRNA coding for tyrosine hydroxylase (TH) increased in the locus coeruleus during 72-h REM sleep deprivation, presumably as a response to the increased neuronal activity and release of noradrenaline.[5] TH mRNA was not affected by a short total sleep deprivation.[6]

This chapter introduces a sensitive and a very low-background assay to detect mRNA using *in situ* hybridization. The method is a modification of that originally published by Miller et al.[7] Each step is described in detail and preceded by a discussion of the particular problems and choices of that step. Recipes for the solutions are given at the end of the section.

The first steps have to be RNase-free to protect the tissue RNA from degradation. RNases are stable enzymes and ubiquitous in biological material. To prevent RNase contamination, the following steps should be taken.

1. The workers should wear gloves (nonsterile, latex or vinyl examination gloves are sufficient) upon all contact with tissue and reagents that will come in contact with the tissue.
2. After being washed, all dishes are wrapped in aluminum foil, and autoclaved or heated to at least 120°C over night.
3. All solutions are made with DEPC (diethyl-pyrocarbonate)-handled water (recipe included) and autoclaved.
4. It is recommended that the work be performed in a separate area of the laboratory, and fresh bench covers should be introduced every day.

RNase-free procedure is needed in all steps until the hybridization step is completed. Hot washes and subsequent steps can be performed without fear of RNase contamination.

II. Protocol

A. Brain Sectioning

The brain sections are cut from frozen tissue; no fixation is needed before sectioning (in contrast to immunohistochemistry, where the tissues are perfused with fixative before sectioning.) Excessive use of fixative (usually paraformaldehyde) will give high background in dark field microscopy.

RNase-free step

1. Fresh whole brain is frozen on a piece of foil placed on dry ice in an insulated container and stored at –70°C until used (storage up to 2 to 3 months is usually safe; in some cases tissues can be stored up to 6 months).
2. The microtome chamber temperature is set at –20 to –23°C. The area of interest is cut from the brain using a razor blade and mounted on a chilled microtome chuck using mounting medium, e.g., Tissue Tec (Miles Inc., Elkhart, IN). The temperature is allowed to equilibrate for at least 15 min before sectioning is started.
3. A model "C" blade is used, with a cutting angle of 15°. 20 μm sections are cut and thaw-mounted on cold (–20°C) gelatin subbed slides. (See separate recipe for preparing the slides.)
4. Slides are stored at –70°C in a slide box (holds 25 slides) with a desiccant capsule and sealed with PVC tape. More than 1 month storage time is not recommended.

B. Preparation Tissue for Hybridization — Postfixation, Acetylation, and Delipidation

Mild postfixation is needed to fix the macromolecules of the tissue on the slide. We use 5 min in 4% paraformaldehyde. To guarantee the potency of the fixative, the solution is made not longer than one day before use, because paraformaldehyde polymerizes quickly in water solutions.

To reduce unspecific binding of the probe, the positively charged molecular groups in the tissue are acetylated by acetic anhydride. Fresh anhydride is introduced to every tray (30 slides), while the reagents in other steps are changed after every 5 trays. This step is finished by dehydrating the tissue through a series of alcohols in preparation for delipidation by chloroform. Delipidation reduces background binding caused by myelin in the white matter.

RNase-free step

1. The slides are removed from the freezer and laid on a sheet of foil for 10 min or until thoroughly dry. After drying they are loaded on a metal slide tray (for 30 slides).
2. Postfixation is performed using 4% paraformaldehyde (made fresh from powder, see recipe) in phosphate buffer pH 7.4 in an ice bath for 5 min. During fixation the slides are agitated gently a couple of times. After fixation the slides are rinsed first in PBS-buffer (pH 7.4) for 2 min in ice bath, followed by rinse in 1.5% TEA-solution (pH 8, see recipe for making TEA) for 10 to 20 sec.
3. Slides are then acetylated using 0.3% acetic anhydride in 1.5% TEA (875 μl acetic anhydride in 350 ml of 1.5% TEA) for 10 min. During this step the solution is agitated with a stir bar. Solution is changed between every tray.
4. Slides are then rinsed in 2X SSC (see recipe for making SSC and its dilutions) for 10 to 20 sec and dehydrated in 70%, 95%, and 100% ethanol be letting them stand 2 min in each solution.
5. Sections are then delipidated in chloroform for 5 min and rehydrated in 100% and 95% ethanol for 2 min each. Slides are dried on trays at room temperature and can be stored at this point for a maximum of 1 week.

C. Preparation of ^{35}S-labeled Oligonucleotide Probe

The antisense probe can be either a long sequence DNA or RNA (usually 100 to 300 bases) made by cloning or a short sequence of DNA (oligonucleotide probe, usually 20 to 60 bases) made artificially by an oligonucleotide synthesizer. This chapter discusses the use of oligonucleotide probes. For this method we have to know the exact base sequence of the messenger RNA. This information is usually obtained from gene databases (e.g., GCG, Entrez). The base sequence is chosen so that the number of guanine (G) and cytosine (C) bases is from 55 to 70% of the total number of bases. The higher the GC content is, the more hydrogen bonds are formed between the probe and mRNA, and the more stable is the binding. If the probe is meant to target a species in which the exact order of bases in not known, we compare corresponding nucleotide sequences between species (usually mouse, rat, and human) whose sequence of the mRNA is known, and we select the part where the base sequence is the same or as similar as possible between the species (conserved sequence). There is a good probability that this nucleotide sequence will work also in species whose sequence of the corresponding mRNA is unknown.

The probe can be labeled by an enzyme (e.g., digoxygenin) or by a radioactive isotope. Currently, the most commonly used radioactive isotopes are ^{32}P and ^{35}S. Both are β^--emitters. The energy of ^{32}P is higher (1.7 MeV) than that of ^{35}S (0.17 MeV), and thus more penetrating. The exposure time is shorter when ^{32}P is used, but ^{35}S gives better resolution. Sulphur groups in the ^{35}S-labeled probes have a tendency to form disulfide bonds with each other and with proteins in the tissue. This can be reduced by adding dithiotreitol (DTT) to the probe buffer. We have used ^{35}S-labeled dATP which is added to the 3′ end of the oligonucleotide by using terminal deoxynucleotide transferase (TdT) enzyme. Proper handling and disposal of radioactive material should be kept in mind, if radioactive labeling is used.

RNase-free step

TdT labeling

The oligonucleotide is labeled at the 3′ end with ^{35}S-dATP using terminal deoxynucleotidyl transferase (TdT) enzyme (Pharmacia Biotech: 27-0730-01). The ^{35}S-dATP and the probe are mixed in a dATP:probe = 5:1 molar ratio.

For one reaction:

Probe	15 pmol
5X enzyme buffer (in TdT kit)	5 μl
^{35}S-dATP	75 pmol
TdT	20 U
Sterile H_2O	add 25 μl

In practice, we have diluted the probe first to yield 10 pmol/μl stock concentration (stored at –20°C), which is diluted to 1.5 pmol/μl for use in hybridizations.

We mix together for one reaction:

Probe (1.5 pmol/μl)	10 μl
^{35}S-dATP (10 mCi/ml, 1 MCi/mmol)	8 μl
5X enzyme buffer	5 μl
TdT (10 U/μl)	2 μl
Total volume	25 μl

Note: *Do not use DEPC-H_2O for labeling, but autoclaved ordinary water instead.*

TdT must be kept on ice until the labeling reaction is started by mixing the four reagents.

The reagents are pipetted into a small RNase-free Eppendorf vial and tapped gently to mix. One half to two reactions can be mixed in one vial. The vial is incubated at 37°C for 1.5 h, and the reaction is stopped by heating to 65°C for 5 min. The labeled oligonucleotide probe is then purified from the reaction mixture.

Purification of the probe

Purification is done across an NENsorb 20 Nucleic Acid Purification Cartridge (NEN #NPL-022, Du Pont Company, Wilmington, DE) according to the following protocol:

Column preparation:

The solution is loaded at the top of the column; the cap piece is then attached to the column, and the solution is pushed through the column using pressure created by an ordinary 10 ml syringe. When the solution is almost through, the cap is removed and the next solution is loaded and pushed through, and so on. Avoid pushing air into the column.

1. 2 ml methanol (HPLC grade)
2. 2 ml 0.1 *M* Tris, pH 8.0
3. 2 ml 0.1 *M* Tris, pH 8.0

Loading of the probe. Start to collect 1 ml fractions:	***Fraction:***
1. Load the probe by flushing the probe vial with 200 μl 0.1 *M* Tris	1
2. Flush the probe vial again with 200 μl 0.1 *M* Tris	1
3. 2 ml 0.1 *M* Tris	1 to 2
4. 2 ml DEPC-H_2O	3 to 4

Sample elution (0.5 ml fractions):	***Fraction:***
5. 0.5 ml 50% n-propanol (in DEPC-H_2O)	5
6. 0.5 ml 50% n-propanol	6
7. Empty the column by pressing air through it	6

8. Take 1 μl sample of every fraction and count for radioactivity. If the reaction and purification have been successful, most of the activity should be in fraction 5.
9. Measure the volume of fraction 5 (= the probe). The probe can be stored. If not used on the same day, add 1% 1 *M* DTT and store at –70°C.

Note: *It is essential to keep track of the activity of the fractions. If a major portion of the activity appears in the wrong fraction or if a considerable amount of activity is lost during later mixings, it is not advisable to proceed, but rather start a new labeling reaction.*

Preparation of the hybridization buffer + probe

1. Mix together:

Probe	3 pmol per 1 ml of the final hybridization mixture
tRNA (10 mg/ml)	25% of the volume of the probe-tRNA
TED-buffer	balance so that the volume of probe + tRNA is 20% of the final volume of the hybridization mixture

2. Heat at 70°C in a water bath for 3 min
3. Cool immediately in ice water
4. Take a 1 μl sample for counting
5. Add hybridization buffer (20% probe-tRNA, 80% buffer), vortex vigorously.
6. Take a 5 μl sample and count with the probe-tRNA sample. This is done to determine the specific activity of the hybridization mix.

D. Hybridization

RNase-free step

1. Add hybridization mix (room temperature) to the slides (45 μl for 24 × 50 mm, 55 μl for 24 × 60 mm) and lay a silanized coverslip on the slides carefully. Avoid air bubbles.
2. The slides are laid on a tray that is covered by moist (DEPC-water) paper towels.
3. Cover the tray with foil and incubate overnight at 37°C. (Moist incubator is recommended if available.)

E. Post Hybridization ("Hot Wash")

The hot wash temperature is one of the critical parameters of the *in situ* hybridization assay. If it is too high, the signal is washed away; if too low, the background will be high. For our probes (48 bases) and the consistency of the hybridization mix, 55°C has proved to be suitable. However, every time a new probe is introduced, it would be advisable to make a test assay, and test the hot wash temperature using increments of 5°C.

RNase-free procedure is not required in this step. The coverslips are washed away in 1X SSC and the slides are reloaded in the metal trays that stand in a 1X SSC solution.

The slides are washed 4 times; each for 15 min in 55°C 1X SSC (water bath shaker). The SSC is changed between each wash (keep 1X SSC supply at the correct temperature on a hot plate or in a water bath). The temperature should be kept between 54 and 56°C. We use a thermometer inside the slide container to monitor the temperature during hot washes. The hot washes are followed by two washes at room temperature (1X SSC), 1 h each (shaker or stirring bar).

Slides are then dehydrated in 70% and 95% ethanol containing 0.3 *M* ammonium acetate and in 100% ethanol for 20 sec in each solution. Let dry at room temperature.

Note: *According to our measurements, the first cold rinsing liquids are mildly radioactive, and they have to be disposed of as radioactive waste. The radioactivity of all other liquids (including liquids from the first hot wash) is not distinguishable from the activity of the background.*

F. Visualization of the Signal Using Photographic Emulsion Coating

Visualization of the signal can be performed either using autoradiographic film or photographic emulsion. In the former, the film is simply placed over the sections for exposure and developed. The signal is read and quantitated from the film. While the quantitation of the signal is as good as from emulsion detection, the resolution is much poorer.

In the emulsion method the sections (on slides) are covered by photographic emulsion, and the signal is exposed into the emulsion layer on the sections. After developing and removal of excess emulsion, the signal can be detected at the cellular level using a microscope (Figure 14.2). For this reason, we use the emulsion coating technique. The disadvantage is that it is both labor-intensive and expensive.

Densitometric analysis of the amount of mRNA in a tissue is easier and more accurate on film than in emulsion, because the emulsion layer is more even in the

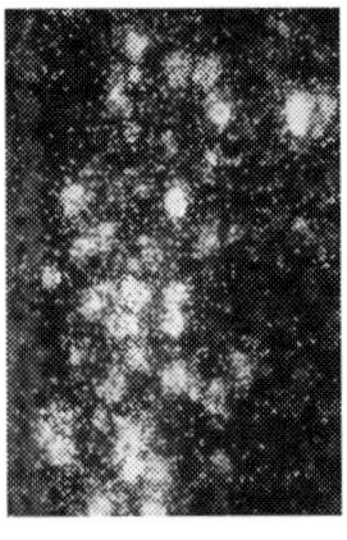

FIGURE 14.2
Photomicrograph of hypothalamic neurons visualized in dark field microscopy after *in situ* hybridization with 48mer ^{35}S oligonucleotide probe for rat somatostatin mRNA.

film and there is no background caused by underlying tissue. In the film the absolute amount of the mRNA can be estimated by using tissue homogenate standards. If resolution of individual mRNA-containing cells in the tissue is needed, e.g., for cell counting, comparison of the histological image and distribution of the signal is possible simultaneously when emulsion is used. In emulsion, relative estimation of the amount of mRNA in the cells or tissue is also possible within an assay. The signal from the tissue structure has to be separated from the signal of the probe, and every measured point must thus have its own background control measurement, which is subtracted from the former. These processes can be achieved with a computer-aided image analysis system.

Before emulsion coating we check the result of the washes using film autoradiography: a few slides are exposed overnight on autoradiography film (without intensifying screens), and the film is developed. If the wash has been effective, the film should be almost clear. If the background is high, the slides can be rewashed at a higher temperature.

An absolutely light-proof darkroom equipped with a dim red safe light (Kodak Safelight filter # 2) is needed. Often the door used as entry to the darkroom leaks light. This can be easily checked by staying inside the darkroom with lights off for 15 min. If light leaks are found, they should be blocked before handling the emulsion. The tiniest light leak will increase the background on the slides.

The emulsion is melted in a 42°C water bath for 15 min. Higher temperatures will increase the background. The emulsion is diluted 1:1 with 0.6 *M* ammonium acetate by pouring it into a decanter and stirring gently. This solution can be stored in 20 ml scintillation vials wrapped in foil and kept at 4°C (avoid freezing) for at least 3 months.

The dipping chamber is filled with diluted emulsion, and the slides are dipped one at a time by entering and exiting the emulsion 2 times with an even, fairly slow speed. The slides are left to dry on a tray, e.g., a cardboard scintillation tube tray, for 2 h at room temperature in total darkness (with the red working light turned off). The slides are packed into boxes with a desiccant capsule, and the boxes are sealed with PVC tape and wrapped in foil to prevent any entry of light. A few slides are put in a separate box for use as test slides. The boxes are stored in a refrigerator preferably reserved for this purpose only. Any radioactive material will increase the background of the slides.

Test slides are developed at 1- to 2-week intervals to evaluate the development of the signal.

1. Development

In the dark room with safelight illumination:

1. Load the slides in metal trays.
2. Develop for 4 min at 16°C in Kodak D-19 developer diluted 1:1 with water. Agitate gently.
3. Rinse in 16°C water for 1 min.
4. Fix in Kodak T-max fixer at 16°C for 5 min.

5. Rinse 30 min in running water or in a container, changing the water frequently. After the fixer, the slides can be handled in daylight.

Fixer can be reused a couple of times, but the developer oxidizes quickly. Fresh developer is colorless or pale brown. A yellow color is the mark of oxidized developer.

2. Counter-Staining

For dark field microscopy, the slides are counterstained lightly by cresyl violet (other stains can also be used).

1. Rinse in 0.1 *M* sodium acetate buffer pH 3.5 for about 1 min.
2. Stain by dipping the slides 3 times in 0.1% cresyl violet in 0.1 *M* sodium acetate. When the sections are lightly stained, the staining is sufficient.
3. Rinse in dH_2O for 20 sec.
4. Dehydrate in 70%, 95%, and 100% ethanol for 20 sec in each liquid.
5. Put the slides in toluene or xylene (or xylene substitute, e.g., Histoclear). If toluene is used, this should be done in a vacuum chamber. (Toluene is neurotoxic and carcinogenic.)
6. Mount the coverslips on the slides. Put 4 drops of mounting medium (e.g., Diatex diluted with toluene 2:1) and gently lay a 24 × 60 mm coverslip on the slide, avoiding air bubbles.
7. Dry overnight in a vacuum chamber.
8. Remove the excess emulsion by wiping with a soft cotton cloth moistened by ammonia-containing multipurpose cleaner (e.g., Ajax, Formula 409). Clean the slide by wiping with a dry cloth.

Reagents Needed

20X SSC pH 7.0, 1 liter

NaCl 175.3 g
Na-Citrate 88.1 g
Add approximately 900 ml ddH_2O, adjust pH to 7.0, bring to volume. Autoclave for 30 min. Dilute with DEPC-H_2O to get 2X SSC (1:10) or 1X SSC (1:20).

PBS pH 7.4, 1 liter

Na_2HPO_4 14.2 g
NaCl 8.8 g
Add 1 liter ddH_2O, adjust pH to 7.4 with HCl.
Autoclave for 30 min.

DEPC-H_2O

DEPC (Diethylpyrocarbonate, Sigma # D5758) 1 ml/1 dH_2O
Shake vigorously, let stand overnight, autoclave for 30 min.

Note: *DEPC is toxic; store in refrigerator.*

4% PFA (paraformaldehyde) pH 7.4, 1 liter

NaH_2PO_4 2.28 g
Na_2HPO_4 11.5 g
Add 800 ml DEPC-H_2O, heat to 58 to 62°C, slowly add 40 g PFA, add 7 pellets of NaOH until clear, adjust to pH 7.4 with HCl, bring to volume, filter with Whatman 2v filter. Work under a hood.

1.5% TEA (triethanolamine) pH 8.0, 1 liter

TEA 15 g
Add 1 liter DEPC-H_2O, adjust pH to 8.0, make fresh.

1 M DTT (dithiothreitol)

DTT 3.08 g
Add 20 ml 0.5 *M* EDTA to a centrifuge tube, divide aliquots into Eppendorf vials, 500 µl each, store at –70°C.

TED buffer (Tris-EDTA-DTT), 1 ml

1 *M* Tris pH 8.0	10 µl
0.5 *M* EDTA pH 8.0	2 µl
1 *M* DTT	10 µl
DEPC-H_2O	add 1 ml

Make fresh.

50% Dextran Sulphate

Dextran sulphate 10 g
Add DEPC-H_2O up to 20 ml in graduated centrifuge tube, vortex, store refrigerated.

Hybridization Buffer

Formamide	0.5 ml	2.5 ml	(50%)
50% dextran	0.2 ml	1.0 ml	(10%)
5 *M* NaCl	60 µl	0.3 ml	(0.3 *M*)
1 *M* Tris pH 8	8 µl	40 µl	(10 m*M*)

0.5 *M* EDTA	1.6 μl	8 μl	(1 m*M*)
50X Denhardt's	20 μl	0.1 ml	(1X)
1 *M* DTT	8 μl	40 μl	(10 m*M*)
DEPC-H_2O	2.4 μl	12 μl	
Buffer volume	0.8 ml	4 ml	
Probe + tRNA	0.2 ml	1 ml	
Total Volume	**1 ml**	**5 ml**	

Subbed Slides

1. Load slides in metal carriers. Secure with rubber bands and wash in dishwasher.
2. Rinse in 1% HCl in 95% ethanol followed by a 95% ethanol rinse. Rinse in running dH_2O for approximately 20 to 30 min (in a bucket).
3. Bake overnight in 160°C oven. Let cool.
4. **(RNase-free)** Dip the slide carriers two times in subbing solution (recipe below) with an even, slow speed avoiding the formation of foam. The foam can be peeled gently from the surface of the solution.
5. Dry in incubator on sheets of foil. Tilt the trays so that the frosted ends of the slides are lower.

Subbing Solution, 1 liter (RNase-free)

Gelatin (Sigma: G-2500)	5 g
Chromium potassium sulphate (Sigma: C-5926)	0.5 g

Dissolve in 1 liter DEPC-H_2O at 37°C.

III. Discussion

We have used this method to detect mRNA changes during and after sleep deprivation. mRNA synthesis precedes protein synthesis, and thus to a certain extent describes the synthetic capacity of the cell.[8] However, a direct correlation between the amount of mRNA and the protein synthesis cannot be made, and thus any statement of increased protein synthesis should be complemented by measurement of the actual protein levels.

The specificity and easy synthesis of oligonucleotide probes have made *in situ* hybridization a tool to localize cells that contain certain proteins, e.g., enzymes or receptors, comparable to immunohistochemistry. The disadvantage is that if the mRNA of the protein is not expressed at the time of the tissue collection, and the half-life of the mRNA is short, no mRNA will be detectable, even if the cell would synthesize this protein. This is again a reason to complement *in situ* hybridization with other methods.

The quantitation of the mRNA message is possible, but can be done within one assay only; it is a semiquantitative method. Quantitation from either film or from emulsion-coated slides is equally possible and accurate; quantitation is done using computerized image analysis systems[9] and consists of reading the original signal, either from film or the emulsion-handled section, to the computer program, analyzing the signal, and calculating it.

In situ hybridization is less sensitive than PCR, another powerful method for detecting mRNA. PCR can be made quantitative,[10,11] but since the tissue for PCR analysis is homogenized, it will not provide an exact localization of the cells that contain the mRNA. If localization is an important part of the information, *in situ* hybridization should be used instead to detect mRNA.

References

1. Toppila, J., Asikainen, M., Alanko, L., Turek, F. W., Stenberg, D., and Porkka-Heiskanen, T., The effect of REM sleep deprivation on somatostatin and growth hormone-releasing hormone gene expression in the rat hypothalamus, *J. Sleep Res.,* 5 (Suppl. 1), 115, 1996.
2. Toppila, J., Asikainen, M., Alanko, L., Ward, D. J., Tobler, I., Stenberg, D., and Porkka-Heiskanen, T., Sleep deprivation by gentle handling increases somatostatin gene expression in the rat, *J. Sleep Res.,* 5 (Suppl. 1), 229, 1996.
3. Toppila, J., Stenberg, D., Alanko, L., Asikainen, M., Urban, J. H., Turek, F. W., and Porkka-Heiskanen, T., REM sleep deprivation induces galanin gene expression in the rat brain, *Neurosci. Lett.,* 183, 171, 1995.
4. Toppila, J., Alanko, L., Asikainen, M., Tobler, I., Stenberg, D., and Porkka-Heiskanan, T., Sleep deprivation by gentle handling does not affect galanin mRNA expression in the rat anterior hypothalamus, *Soc. Neurosci. Abstr.,* 22, 146, 1996.
5. Alanko, L., Toppila, J., Asikainen, M., Ward, D. J., Tobler, I., Stenberg, D., and Porkka-Heiskanen, T., Tyrosine hydroxylase gene expression in the locus coeruleus is not affected by total sleep deprivation in the rat, *J. Sleep Res.,* 5 (Suppl. 1), 3, 1996.
6. Porkka-Heiskanen, T., Smith, S. E., Taira, T., Urban, J. H., Levine, J. E., Turek, F. W., and Stenberg, D., Noradrenergic activity in the rat brain during REM sleep deprivation and rebound sleep, *Am. J. Physiol.,* 268, R1456, 1995.
7. Miller, M. A., Kolb, P. E., and Raskind, M. A., Testosterone regulates galanin gene expression in the bed nucleus of the stria terminalis, *Brain Res.,* 611, 338, 1993.
8. Debure, L. I., Moyse, E., Fevre-Montange, M., Hardin, H., Belin, M. F., Rousset, C., Pujol, J. F., and Weissmann, D., Somatotopic organization of tyrosine hydroxylase expression in the rat locus-coeruleus — long term effect of RU24722, *Brain Res.,* 581, 19, 1992.
9. Gerfen, C. R., Quantification of in situ hybridization histochemistry for analysis of brain function, in *Gene Probes,* Conn, P. M., Ed., Academic Press, San Diego, 1989, Chap. 5, 79.
10. Dallman, M. J. and Porter, A. C. G., Semiquantitative PCR for the analysis of gene expression, in *PCR. A Practical Approach,* McPherson, M. J., Quirke, P., and Taylor, G. R., Eds., Oxford University Press, Oxford, 1991, Chap. 13, 215.
11. Ikonen, E., Manninen, T., Peltonen, L., and Syvänen, A. C., Quantitative determination of rare mRNA species by PCR and solid-phase minisequencing, *PCR Methods & Applications,* 1, 234, 1992.

Chapter 15

New Directions in the Analysis of Brain Substances Related to Sleep and Wakefulness

Gary Siuzdak and Steven Henriksen

Contents

I. A Brief History of CSF Characterization

Between 1916 and 1960 cerebrospinal fluid (CSF) was believed to be little more than an ideal physiological saline solution, unidentifiable by any known histochemical method.[1] In the 1970s, characterization of CSF began in earnest and as the technology for its analysis developed, much was learned about the cation, anion, and nonelectrolyte (specifically urea and glucose) content of CSF. The primary

0-8493-3361-X/98/$0.00+$.50

species identified therein were Na^+, K^+, Mg^{2+}, Ca^{2+}, Cl^-, HCO_3^-, $^{82}Br^-$, $^{131}I^-$, CNS^-, glucose, and urea. During this period it was also hypothesized that CSF had additional complexity relating to its interactions with blood plasma across the blood–brain barrier.

In the 1980s, significant discoveries within CSF included sphingolipids, brain-specific proteins, nucleic acids, transfer RNAs, and amino acids. Beyond discovering the mere presence of these molecules, their detailed characterization was possible through advances in analytical technology, namely gas chromatography mass spectrometry, DNA sequencing, and amino acid sequencing. Substances which had been previously overlooked, like myelin basic protein, S-100 protein, NP (ribonucleoprotein), 14-3-2 protein, antigen-α, and α_2-glycoprotein, could now be completely characterized.

II. Gaps in Existing Knowledge

The increased knowledge of the content of cerebrospinal fluid has brought with it a vast array of new determinants for the presence and course of diseases. By gaining a better understanding of CSF, physicians and researchers alike have been able to better detect and monitor treatment effectiveness for diseases such as Alzheimer's, AIDS-related neurological disease, muscular sclerosis, and tuberculosis meningoencephalitis. Unfortunately, it is clear that the current reference files for CSF are far from complete. While physicians are relying more and more on clinical chemistry laboratories for biochemical analysis of patients' CSF, the clinical value of many CSF analytes outside of glucose and protein determinations is often ambiguous. Thus, correlating existing analytes and identifying new analytes that correlate with specific physiological processes has immediate utility. Much of the current work on CSF is still directed toward finding the process or pathway of exchange and transport of the CSF from the brain to the locations in which its constituents are utilized. Still lacking is complete knowledge of its actual content and the relationship of these compounds with physiological processes. The goal in current research lies in identifying these molecules primarily through new analytical techniques.

III. Filling the Gap

Our group has taken a purely chemical approach in its attempt to identify molecules of the central nervous system which vary according to various physiological stimuli. Of particular interest has been the identification of a compound or compounds correlating to sleep.[2] A driving force for these experiments was the realization that the sensitivity of modern analytical techniques could allow for a fresh look at CSF constituents.[3-5] Most notably, the sensitivity offered by mass spectrometry has provided a route by which picomole to femtomole amounts of a compound may be analyzed, whereby thermally labile biomolecules are less apt to fragment and decompose than they were with earlier mass analysis techniques such as electron ionization.[6] The goal of our experiments has been to use this technology to identify new molecules in CSF associated with the sleep–wake cycle.

A. Mass Spectrometry

The attempt to identify sleep molecules in CSF requires an efficient method for screening its molecular content. Electrospray ionization (ESI) and matrix-assisted laser desorption/ionization (MALDI) mass spectrometry both offer molecular weight information on these types of compounds. In fact, it has been the development of these mass analysis techniques, allowing higher levels of sensitivity, increased mass range, and better mass accuracy, along with new sampling methods, that have led to an increasing number of mass spectrometry-based applications.[6,7]

ESI and MALDI are fundamentally very different ionization techniques, yet they achieve essentially the same result — the generation of gas phase ions via nondestructive vaporization and ionization. In both techniques, ionization typically occurs through proton addition or proton abstraction to produce either $[M+H]^+$ or $[M-H]^-$ ions (where M is the molecule of interest). In matrix-assisted laser desorption/ionization, gas phase ions are generated by the laser vaporization of a solid matrix/analyte mixture in which the matrix (usually a small crystalline organic compound) strongly absorbs the laser radiation and acts as a receptacle for energy deposition.[8] This concentrated energy deposition results in the vaporization and ionization of both matrix and analyte ions containing very few charges. The relatively low number of charge states observed in MALDI makes the technique especially well suited for the analysis of multicomponent mixtures because individual components can be easily identified by the signal generated from their 1+ charge state. This ability to analyze heterogeneous samples has made it especially useful for looking at biological solutions. While not used extensively in this study, MALDI has been useful for the analysis of biopolymers in CSF,[9] and has become the method of choice for large biopolymer (>30 KDa) analysis. ESI, on the other hand, has become more important for analyzing smaller biomolecules, and was therefore our method of choice.

In ESI-MS, ions are formed directly from solution (usually an aqueous or aqueous/organic solvent system) by creating a fine spray of highly charged droplets in the presence of a strong electric field (Figure 15.1). Subsequent vaporization of these charged droplets results in the production of charged gaseous ions. The number of charges retained by an analyte can depend on such factors as the composition and pH of the electrosprayed solvent as well as the chemical nature of the sample. For large molecules, the ESI process typically gives rise to a series of multiply charged species for a given analyte. Because mass spectrometers measure the mass-to-charge (*m/z*) ratio, the resultant ESI mass spectrum contains multiple peaks corresponding to the different charged states (Figure 15.1).

An important feature of ESI-MS is its ability to directly analyze compounds from aqueous or aqueous/organic solutions, a feature that has established the technique as a convenient mass detector for liquid chromatography (LC). ESI also allows for MS analysis at relatively high LC flow rates (1.0 ml/min) and high mass accuracy

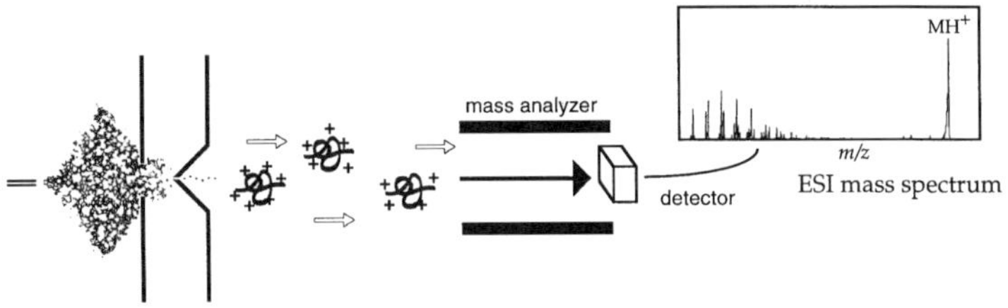

FIGURE 15.1
Schematic representation of electrospray ionization (ESI). ESI occurs via charged droplet evaporation, and the ionized species are directed through a mass analyzer which allows for differentiation and detection of the ions according to their mass-to-charge ratio (*m/z*).

(±0.01%), adding a new dimension to the capabilities of LC characterization. In fact, using ESI-MS as a detector for LC was one of its first obvious applications. Numerous reports have been published on this with special emphasis on small molecule, peptide, and protein analysis. The combination of LC and ESI-MS is excellent for routine and reproducible molecular weight determinations in a wide variety of compounds, whether they are positively (i.e., peptides and proteins) or negatively (i.e., oligonucleotides) charged (Figure 15.2). This is especially useful since the CSF contains both cationic and anionic compounds.

B. Tandem Mass Spectrometry

While molecular weight information is of course useful in the preliminary stages of molecular characterization, it is also important to gather structural information. Tandem mass spectrometry, with its ability to induce fragmentation and perform successive mass spectrometry experiments on these fragment ions, is generally used to obtain this structural information (abbreviated MS^n — where **n** refers to the number of generations of fragment ions being analyzed). The technique is illustrated in Scheme 15.1.

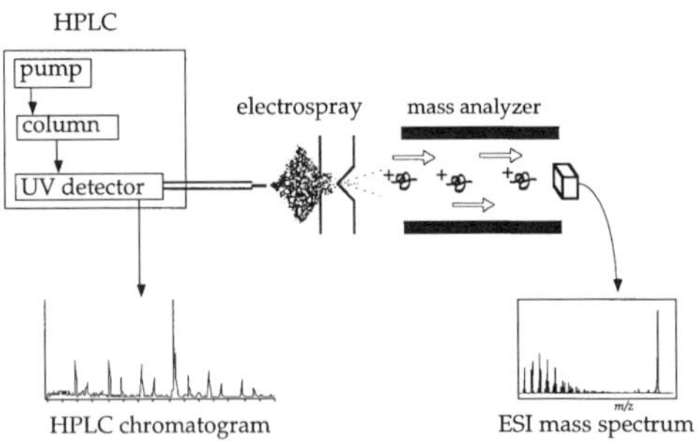

FIGURE 15.2
Interfacing liquid chromatography with electrospray ionization mass spectrometry.

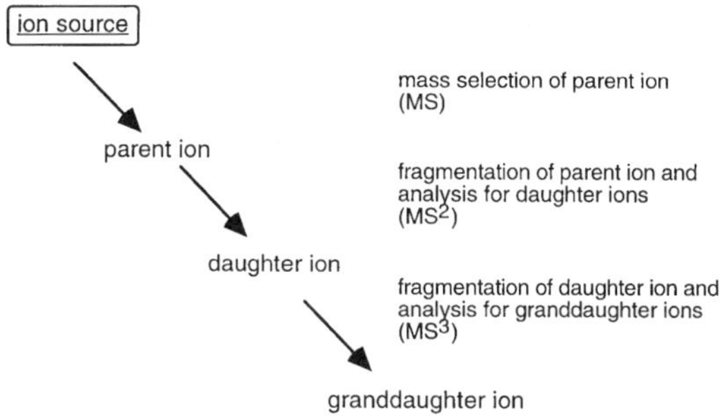

SCHEME 15.1

Fragmentation is usually achieved by inducing ion/molecule collisions. This process, known as collision-induced dissociation (CID), is accomplished by selecting an ion of interest with the mass analyzer and introducing that ion into a collision cell. The selected ion will hit a collision gas, such as argon, which may result in fragmentation. The fragments can then be analyzed to obtain a daughter ion spectrum. ESI typically uses a triple quadrupole or an ion trap mass spectrometer with collision-induced dissociation to perform these analyses. A brief description of MS^2 experiments with a triple quadrupole mass analyzer is given in Figure 15.3.

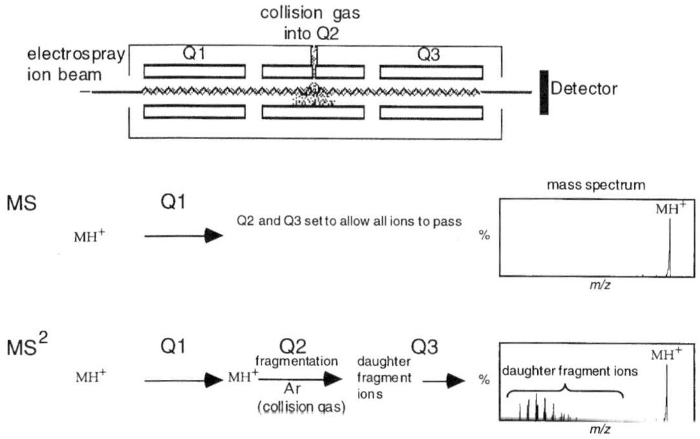

FIGURE 15.3

A triple quadrupole ESI mass spectrometer with ion selection and fragmentation capabilities. Each quadrupole has a separate function: The first quadrupole (Q1) is used to scan across a preset *m/z* range or to select an ion of interest; the second quadrupole (Q2), also known as the collision cell, transmits the ions while introducing a collision gas (argon) into the flight path of the ion selected by Q1; and the third quadrupole (Q3) serves to analyze the fragment ions generated in the collision cell (Q2). For an MS experiment Q1 scans over a selected *m/z* range and all the ions are observed. In the MS^2 experiment, the molecular ion M^+ can be selected by Q1, which results in its fragmentation at Q2. Analysis of the fragments occurs at Q3.

IV. Applications to CSF Studies

In the sleep/CSF studies, fragmentation data were extremely helpful because preliminary analysis of the observed compounds allowed us to gain some characterization information. For example, Figure 15.3 illustrates a fragmentation pattern that allowed us to identify the 22 carbon lipid amide, euricamide. Although this particular compound has not exhibited any sleep-inducing properties in rats, its identification was interesting nonetheless and was one of many results obtained directly from our mass spectral analysis.

With the goal of identifying new molecules associated with the sleep–wake cycle, the sleep–wake states of feline subjects were examined. CSF analysis began with preparative liquid chromatography fraction collection in which UV data were used to determine any differences between the felines' CSF at various points in their sleep cycle. The LC analysis of the felines' CSF produced a peak which was present in sleep-deprived, but not in normal cats. An absorbance in the LC chromatogram was found to be particularly prominent in the CSF of cats that were kept awake for an extended period of time (18 hours). The fractions isolated by LC were then analyzed by electrospray mass spectrometry and electrospray tandem mass spectrometry. Even though the compound associated with this absorbance peak was only present in small amounts, partial characterization was initially obtained by performing electrospray tandem mass analysis and exact mass measurements.

A. Identification

Electrospray mass analysis on the fractions associated with the differences in the chromatogram produced a significant ion at MH^+ *m/z* 282, while exact mass determination on the unknown compound using fast atom bombardment (FAB) mass analysis provided a molecular formula of $C_{18}H_{35}NO$. Collision-induced dissociation (CID) was used to perform MS^2 and MS^3 experiments on the ion at *m/z* 282, in which tandem mass analysis (Scheme 15.1) at *m/z* 282 revealed a distinct fragmentation pattern in the low molecular mass range, consistent with other long-chain alkanes. Neutral losses of 17 and 35 Da from the parent ion indicated a loss of ammonia followed by water. Performing additional MS^3 experiments on the daughter ions at *m/z* 265 and 247 revealed that the daughter ion at 265 fragmented to form the granddaughter ion at 247. This suggested that the ion at 247 was the result of sequential losses (loss of 17 Da [NH_3] followed by 18 Da [H_2O]), as opposed to a neutral loss independent of the daughter ion at 265. Additional deuterium exchange experiments were consistent with at least two protons on this molecule being exchangable. These results suggested that we had a fatty amide containing one point of unsaturation.

Chemical degradation techniques were first employed on synthetic fatty acid amides, identifying ozonolysis as conducive to the analysis of these agents. GCMS analysis of the ozonolysis reaction mixture derived from the natural lipid revealed nonyl aldehyde as the only C-terminal aldehyde present. Nonyl aldehyde corresponds to an olefin positioned seven methylenes away from the terminal methyl

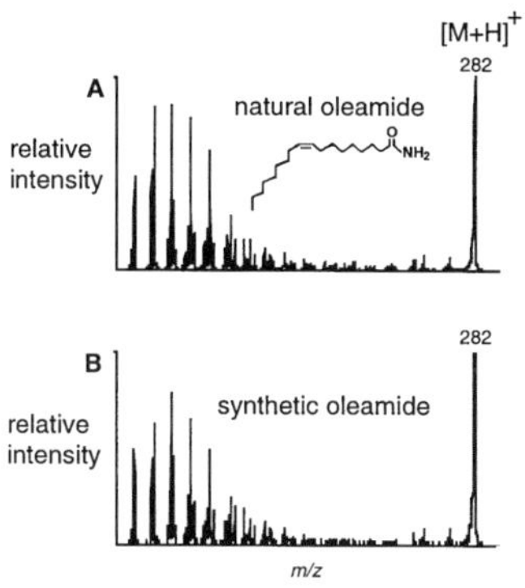

FIGURE 15.4
Electrospray ionization tandem mass spectra of natural oleamide found in CSF (A) and the synthetic version (B).

group of the alkyl chain, which in the case of a C18 fatty acid primary amide, is the 9,10 position.

On the basis of these experiments, compounds that best correlated to the data were synthesized and spectroscopically evaluated. The synthesis of fatty amides, combined with NMR and IR spectroscopic and mass spectral data of these compounds led to the unknown identification as *cis*-9,10-octadecenoamide or oleamide (Figure 15.4). A summary of the data appears in Table 15.1.

Mass Spectrometry. The mass spectrometry experiments were performed on an API III Perkin Elmer SCIEX triple-quadrupole mass spectrometer and a Fisons/VG ZAB-VSE high resolution magnetic sector mass spectrometer. The pneumatically assisted electrospray interface on the SCIEX was used for sample intro-

TABLE 15.1
Summary of Data Generated on Oleamide

Data type	Summary of data
Cationization	Electrospray MS observation of *m/z* 304 $[M + Na]^+$ and *m/z* 320 $[M + K]^+$ confirms that M = 281.
Deuterium exchange	*m/z* = 285 $[M - 2H^+ + 3D^+]^+$, at least 2 exchangeable protons.
Exact mass	Observed 304.2614 Da
	Theoretical 304.2616 Da exact mass for $[C_{18}H_{35}NO + Na]^+$
ESI-MS^2	Fragment ion *m/z* 265 corresponding to $[MH - NH_3]^+$.
	Fragment ion *m/z* 247 corresponding to $[MH - NH_3 - H_2O]^+$.
	Lipid fragment ions.
ESI-MS^3	Fragmentation of *m/z* 265 >> *m/z* 247 (sequential loss) suggests *m/z* 247 fragment is related to *m/z* 265 fragment ion.
	Lipid fragment ions.
NMR data	Confirms lipid portion of molecule and configuration of double bond.
Ozonolysis with GC/MS	Location of the double bond.
IR data	Identities unknown as *cis* isomer.

duction with the potential of the interface sprayer maintained at 5.0 kV. A curtain gas of ultra pure nitrogen (1.0 liter/min.) between the interface plate and the sampling orifice was applied to aid desolvation of the charged droplets and to prevent particulate matter from entering the analyzer region. Samples were introduced through the interface at a rate of 4.0 μl/minute. The positive ions generated by the ion evaporation process entered the analyzer through an interface plate and a 100 μm orifice. The declustering potential was maintained between 50 and 250 V (typically 100 V) to control the collisional energy of the entering ions. A cryogenic pump was used to cool the surfaces within the spectrometer (14 to 18K) maintaining a working pressure of 2×10^{-5} Torr and a sealed pressure of 8×10^{-8} Torr in the analyzer region. The Fisons/VG ZAB-VSE instrument was used to perform the exact mass determination on the unknown compound by fast atom bombardment (FAB) analysis; a resolution of 10,000 (10% valley definition) was obtained to perform an exact mass measurement to within 0.7 parts per million.

Collision-Induced Dissociation (CID). The CID experiments were performed on the API III Perkin Elmer SCIEX triple-quadrupole mass spectrometer with ultra pure argon (>99.99% purity) as a collision gas. The positive ion MS/MS data was acquired by mass selecting the precursor ion with the first quadrupole, after which collisions with argon (target thickness of 3×10^{14} atom/cm^2) in the second quadrupole produced dissociation. The third quadrupole mass-analyzed the resultant daughter ions. Collision energies of 80 eV were maintained in these experiments. CID spectra shown were the result of averaging from 50 to 100 scans depending on the number of scans necessary to obtain a signal-to-noise greater than or equal to 50. MS/MS/MS (MS^3) data were also obtained by mass selecting a daughter ion generated at the orifice with the first quadrupole mass analyzer. Collisions with argon (target thickness of 3×10^{14} atom/cm^2) in the second quadrupole collision cell produced further dissociation, and the resultant granddaughter ions were analyzed with the third quadrupole mass analyzer.

B. Characterizing Physiological Properties of Oleamide

Following its identification, synthetic oleamide was tested on rats. The synthetic compound was injected intraperitoneally into rats in doses of 1 ($n = 2$), 2 ($n = 2$), 5 ($n = 7$), 10 ($n = 10$), 20 ($n = 2$), and 50 ($n = 2$) mg, where n is the number of rats. Two hours after the lights cycled off (in a 12-hour light/12-hour dark cycle) the low doses of 1 and 2 mg produced no overt effect on behavior. Doses of 5 mg or more, however, induced long-lasting motor quiescence, with closed eyes and sedate behavior characteristic of normal sleep. Also as in normal sleep, the rats still responded to auditory stimuli with an orienting reflex and sustained attention toward the source of stimulation. Behavioral sedation began nearly 4 minutes after injection and lasted 1 hour (for 5 mg), 2 hours (10 mg), or 2.5 hours (for both 20 and 50 mg).

Other known effector molecules were found to work in quantities of approximately 2.8 micrograms. When 2.8 micrograms (10 nmol) of oleamide was introduced into two rats intraventricularly, it induced electrophysiologically monitored sleep, indicating that the agent acts directly in the brain at a dose comparable with other known effector molecules.

A vehicle of 5% ethanol in saline solution was used to introduce the synthetic carboxamide into the rats. To ensure that neither the vehicle nor oleic acid produced the observed effect, each was introduced in the rats and no overt behavioral effect or modification of the spontaneous sleep–wake cycle occurred. *Trans*-9,10-octadecenoamide exhibited similar, though reduced, effects compared to those of the *cis* isomer. At 10 mg per rat, the effect was 1 hour of sleep instead of the 2 hours noted for the *cis* isomer. Moving the olefin to the 8,9 or 11,12 positions or extending the alkyl chain length decreased its effect markedly. Although the animals still experienced loss of alertness, their eyes remained open and their alertness was only slightly affected.

Recent reports[11] on oleamide hydrolysis in rat brain membrane fraction observed rapid conversion of oleoamide to oleic acid by rat brain membrane fractions. While amide hydrolysis activity was noted in the rat brain soluble fractions, no appreciable ability to hydrolyze the amide to oleic acid was observed in rat pancreatic microsomes and proteinase K. It is conceivable, given this enzymatic hydrolysis of the compound in the brain and the neutral loss of 17 mass units in the initial spectra, that the liberated ammonia also plays a role in the effector function.

Other fatty amide molecules have been reported recently,[12-14] denoting the existence of a class of molecules in which simple variations of a core chemical structure have distinct physiological consequences. Searching for these molecules and uncovering their physiological role will be the basis for further study as this group continues to probe the CSF using continuously evolving techniques in mass spectrometry.

References

1. Wolstenholme, G. E. W. and O'Connor, C. M., *The Cerebrospinal Fluid, Production, Circulation, and Absorption,* Little, Brown, and Company, Boston, 1958, 124.
2. Cravatt, B. F., Prospero-Garcia, O., Siuzdak, G., Gilula, N. B., Hendriksen, S. J., Boger, D. L., and Lerner, R. A., Chemical characterization of a family of brain lipids that induce sleep, *Science,* 268, 1506, 1995.
3. Kusmierz, J. J. and Desiderio, D. M., Characterization of an aminopeptidase in cerebrospinal fluid — structure elucidation of enzyme hydrolysis products of synthetic methionine-enkephalin by reversed-phase high-performance liquid chromatography and mass spectrometry, *J. Chrom.-Biomed. Applications,* 574, 189, 1992.
4. Eriksson, U., Andren, P., Silberring, J., Nyberg, F., and Wiesel, F. A., Characterization of neurotensin-like immunoreactivity in human cerebrospinal fluid by high-performance liquid chromatography combined with mass spectrometry, *Biological Mass Spectrometry,* 743, 225, 1994.
5. Muck, W. M. and Henion, J. D., Determination of leucine enkephalin and methionine enkephalin in equine cerebrospinal fluid by microbore high-performance liquid chromatography and capillary zone electrophoresis coupled to tandem mass spectrometry, *J. Chrom.-Biomed. Applications,* 495, 41, 1989.

6. Siuzdak, G., *Mass Spectrometry for Biotechnology,* Academic Press, San Diego, 1996, 162.
7. Siuzdak, G., The emergence of mass spectrometry in biochemical research, *Proc. Nat. Acad. Sci. USA,* 91, 11290, 1994.
8. Hillenkamp, F., Karas, M., Beavis, R. C., and Chait, B. T., Matrix-assisted laser desorption ionization mass spectrometry of biopolymers, *Analytical Chemistry,* 63, A1193, 1991.
9. Koudinov, A. R., Koudinova, N. V., Kumar, A., Beavis, R. C., and Ghiso, J., Biochemical characterization of Alzheimer's soluble amyloid beta protein in human cerebrospinal fluid: association with high density lipoproteins, *Biochemical and Biophysical Research Communications,* 223, 592, 1996.
10. Fenn, J. B., Mann, M., Meng, C. K., Wong, S. F., and Whitehouse, C. M., Electrospray ionization-principles and practice, *Mass Spectrom. Rev.,* 9, 37, 1990.
11. Cravatt, B. F., Giang, D. K., Mayfield, S. P., Boger, D. L., Lerner, R. A., and Gilula, N. B., Molecular characterization of an enzyme that degrades neuromodulatory fatty-acid amides, *Nature,* 384, 83, 1996.
12. Facci, L., Dal Toso, R., Romanello, S., Buriani, A., Skaper, S. D., and Leon, A., Mast cells express a peripheral cannabinoid receptor with differential sensitivity to anandamide and palmitoylethanolamide, *Proc. Acad. Sci. USA,* 92, 3376, 1995.
13. Wakamatsu, K., Masaki, T., Itoh, F., Kondo, K., and Sudo, K., Isolation of fatty acid amide as an angiogenic principle from bovine mesentery, *Biochem. Biophys. Res. Comm.,* 168, 423, 1990.
14. Devane, W. A., Hanus, L., Breuer, A., Pertwee, R. G., Stevenson, L. A., Griffin, G., Gibson, D., Mandelbaum, A., Etinger, A., and Mechoulam, R., Isolation and structure of a brain constituent that binds to the cannabinoid receptor, *Science,* 258, 1946, 1992.

Chapter 16

Sleep and Circadian Rest–Activity Rhythms in Prion Protein Knockout Mice

Irene M. Tobler, Marek Fischer, and Jean C. Manson

Contents

I. Introduction

A. The Prion Protein Diseases

One of the major reasons for the numerous gene targeting experiments undertaken in the last few years has been to investigate gene function. The transmissible spongi-

0-8493-3361-X/98/$0.00+$.50

form encephalopathies (TSE) are slow neurological disorders leading to neuronal degeneration in humans and animals.[1] Point mutations of the human prion protein gene can lead to familial forms of prion diseases, such as Creuzfeldt-Jakob's disease, Gerstmann-Sträussler-Scheinker syndrome, and fatal familial insomnia (FFI).[2-4] Scrapies and bovine spongiform encephalitis occur naturally in sheep, goats, and bovines and can be experimentally transmitted to mice and hamsters (reviewed in reference 1). Also the inherited prion diseases, e.g., FFI, have recently been transmitted to mice.[5] Moreover, mice with an amino acid change homologous to the mutation in humans, develop scrapie-like disease.[6] The gene which encodes for the prion protein (PrP) has been assigned to the human chromosome 20[7] and to the corresponding mouse chromosome 2.[8]

B. Normal Function of the Prion Protein

The normal function of PrP, a neuronal membrane protein anchored to the outer surface of neurones, has not yet been established. The PrP gene is expressed at high levels in neuronal cells of the adult brain, and in astrocytes and oligodendrocytes.[9-11] Lower levels of PrP mRNA have been detected in the heart, lung, and spleen.[12] PrP mRNA has also been detected during mouse embryogenesis in the extraembryonic tissue and in the developing central and peripheral nervous system.[13] This suggests a role for PrP in promoting neuronal cell differentiation and in maintaining neuronal function in the differentiated neurons. However, since PrP has also been detected in astrocytes and oligodendrocytes, during kidney development, in the developing tooth bud, and in placenta amnion and yolk sac, the role of PrP may be more widespread, perhaps as part of a cell signaling system required for differentiation of specific cells or as a cell adhesion molecule.

C. Prion Protein Knockout Mice

The aim of the generation of PrP-null mice was twofold: 1) to address the question of the normal function of PrP, and 2) to test the hypothesis that mice devoid of PrP should be resistant to scrapie pathogenesis after infection with mouse scrapie prions.[14-16] The null mice appeared to develop and reproduce normally, and no abnormal behaviors were detected.[14,16] However, PrP-null mice were shown to display weakened long-term potentiation in the hippocampal CA1 subfield, and our results indicated altered regulation of circadian activity rhythms and sleep.[17,18] The circadian abnormalities have been detected in two lines of PrP-null mice. Since the two lines of mice have been produced by different targeting strategies, these results indicate that the phenotypic differences are a result of the loss of PrP and are not an artifact of gene targeting and subsequent mouse production. However, it remains difficult to ascribe the differences between knockout and wild-type mice to a specific function of PrP or to compensatory effects of the organism in the absence of PrP.

D. Fatal Familial Insomnia

Patients afflicted with FFI exhibit a progressive and severe reduction of total sleep time and a disruption of the sleep–wake cycle.[2,4,19] Moreover the 24-h oscillations of hormones related to sleep, such as growth hormone and prolactin, as well as cortisol, ACTH, and melatonin, are disrupted, indicating that the regulation of the circadian system may be disturbed.[20-22] Positron emission tomography showed that the main neurological feature in FFI is the loss of neurons and astrogliosis in the thalamus, particularly in the medio-dorsal and anterio-ventral thalamic nuclei. Other brain regions involved are the frontal cortex, basal ganglia, and limbic areas. The sleep disruption is associated with a reduction and gradual disappearance of spindle activity.

Cellular changes in the thalamic and cortical neurons have been implicated in the regulation of slow waves and EEG spindles.[23,24] GABA synapses may be involved in sleep regulation, and GABA is the principal neurotransmitter in the suprachiasmatic nucleus and the intergeniculate leaflet, important regions for the control of circadian rhythms.[25] GABA receptor-mediated fast inhibition was weakened in PrP-null mice[26] and alterations in GABA metabolism were found in neurons of several brain regions of hamsters infected with scrapie.[27] Moreover, synaptic degeneration was a primary neuropathological feature in prion disease.[28] It is possible that the normal PrP gene product may play a role in circadian and sleep regulation, and that the EEG and sleep pathology in FFI may be related to loss of the normal function.

II. Protocol

A. Outbred and Inbred Prion Protein Knockout Mice

The generation of the mice devoid of PrP that we investigated was based on disrupting one Prn-p allele of embryonic stem (ES) cells derived from agouti 129/Sv (ev) or 129/Ola mice by homologous recombination.[14,16] The wild-type PrP gene has been cloned and sequenced.[29,30] The mouse PrP gene consists of three exons, two of which encode the mRNA 5′ untranslated region and a single 2 kb coding exon (exon III) which comprises the entire coding region, separated by the 5 to 12 kb intron (Figure 16.1). The exons are separated by an upstream intron of 2 kb and a downstream polymorphic intron of approximately 5 to 12 kb.[31] The promoter region has several conserved motifs (e.g., Sp1 and AP-1) as well as other motifs which may serve as sites for further transcription factors. Two different strategies were used to introduce mutations into exon 3 of the PrP gene, both leading to the production of PrP-deficient mice, using a replacement targeting vector with a positive selection marker for DNA integration, the neomycin phosphotransferase gene (*neo*) (Figure 16.1, II; vectors A and B). In one approach, *neo* was inserted under the control of the metallothioneine promoter into the PrP coding sequence, leading to the homologous recombination at the PrP locus without deletion of any PrP sequences (Figure 16.1, II; A[16]). In this approach the herpes simplex virus thymidine

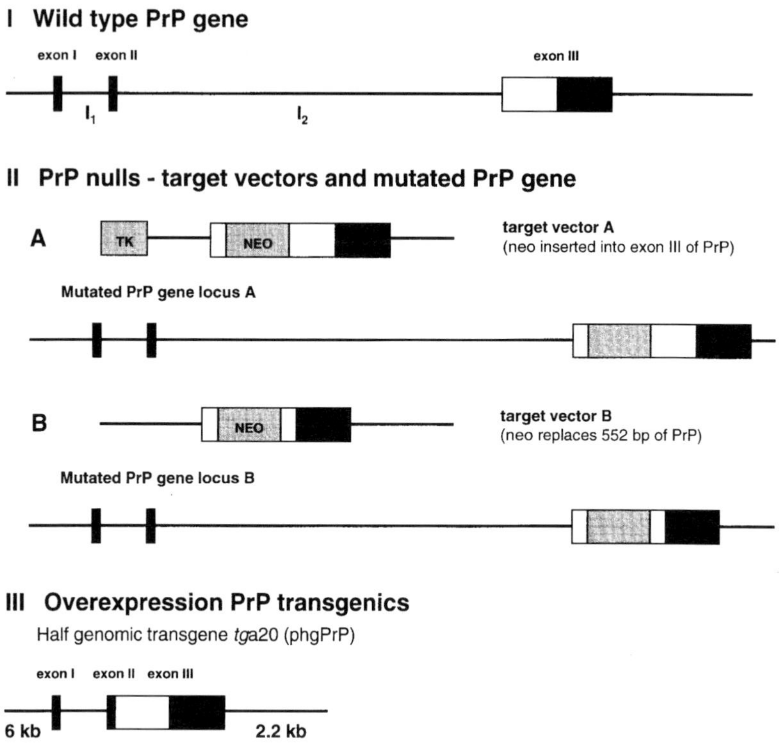

FIGURE 16.1
I. Wild-type PrP gene showing exon (exon I-III) separated by intron 1 (I_1, 2 kb) and 2 (I_2, 6 to 12 kb). Protein coding region (white bar).[31] II. PrP null mice constructed with two different target vectors (A and B). Target vector A[16]: the neomycin phosphotransferase gene (*neo*) under control of the metallothioneine promoter was inserted into exon III of the PrP gene. The thymidine kinase gene (tk) was inserted 5′ of the PrP sequences to select against random integration. The vector was linearized outside the PrP coding region and electroporated into 129/Ola mouse embryonic stem cells. Homologous recombination of this vector with the endogenous PrP locus results in the loss of the tk and a mutated PrP gene (locus A) with the *neo* gene inserted into exon III. No PrP sequences have been removed in this strategy. Target vector B[14]: *Neo* under the control of the thymidine kinase promoter was inserted into exon III of the PrP coding region replacing 552 bp of PrP sequence which had been removed (positions 10-562). The vector was linearized outside the PrP sequence and electroporated into 129/Sv//ev mouse embryonic stem cells. Homologous recombination of the targeting vector with the endogenous PrP locus resulted in a mutated PrP gene locus B lacking 552 bp of PrP and with *neo* inserted into exon III. III. Transgenic mice overexpressing the PrP gene. Map of "half-genomic" construct encoding wild-type PrP gene.[33] Intron 2 has been removed, and only 2.2 kb of the 3′-flanking region remain.

kinase (tk) promoter gene was used as a negative selection against random integration of the DNA. In the second strategy, the targeting vector replaced 552 base pairs of the Prn-p coding region (extending from positions 10 to 562) by a 1.1-kb cassette containing the *neo* gene under the control of tk (Figure 16.1, B[14]).

Both targeting vectors were linearized outside the PrP sequences and electroporated into embryonic stem cells. Targeted clones were selected by PCR and Southern analysis of genomic DNA, confirming that one Prn-p allele was disrupted (frequency of homologous recombination was approximately 1:5000 and 1:800 G418-resistant colonies in the two strategies). The ES cells of one clone were introduced into 4-day-old blastocysts from C57BL/6J mice and implanted into foster mothers, from which chimeric mice were generated. To facilitate breeding, the male chimeras were bred with C57BL/6 females, and screening of the agouti offspring by PCR analysis produced an outbred line of mice which were approximately 50% heterozygous for the disrupted Prn-p allele (confirmed by Southern analysis). A single copy of the targeting sequence was integrated. Breeding of the Prn-$p^{+/0}$ mice gave rise to 24% offspring homozygous for the disrupted Prn-p gene (Prn-$p^{0/0}$; confirmed by Southern analysis). Northern analysis of brain RNA from Prn-$p^{0/0}$ revealed a product of approximately 2.4 kb, which hybridized with both a *neo*-probe and a Prn-p probe comprising the 3′ part of the coding sequence still present in the disrupted gene. However, PrP was undetectable in Prn-$p^{0/0}$ brains and present at about half the normal level in the brains of heterozygous mice (Prn-$p^{0/+}$).

The inbred 129/Ola Prn-$p^{0/0}$ mice were produced by injecting the mutant cell lines into 4-day-old C57BL/CBA mice and the male chimeric mice bred with 129/Ola females. Comparisons of the resulting inbred Prn-$p^{0/0}$ mice with Prn-$p^{+/+}$ allow us, in contrast to the Prn-p null mice with mixed background, to attribute differences in phenotypes to the PrP gene.

B. Overexpression Prion Protein Transgenic Mice

The differences in sleep and circadian rhythms observed in the 129/Sv mice might be due to the undefined, mixed genetic background of the knockout mice.[32] Thus, it was necessary to investigate whether complementation of PrP-null mice with a PrP transgene would restore the normal phenotype, i.e., reduced sleep fragmentation and shortening of the free-running activity period.

The detailed protocol for the generation of a transgenic mouse line where murine Prn-p transgenes were introduced into Prn-$p^{0/0}$ mice is described in Fischer et al.[33] (*tg*a-20 mice). The rationale for constructing this mouse was to investigate whether a reintroduction of normal, wild-type PrP could restore infectivity in null mice.

For the generation of *tg*a-20 mice a PrP-encoding construct was generated comprising exons 2 and 3, flanked by only 2.2 kb of 3′-flanking region, exon 1 and the small first 2 kb intron at the 5′ side, and 6 kb of 5′-flanking region. The large intron was deleted (Figure 16.1; III). Thus *tg*a-20 is a "half-genomic construct" (phgPrP) lacking one intron. This construct was injected into oocytes of Prn-$p^{+/+}$ (C57BL/6J) females, fertilized by Prn-$p^{0/0}$ males. The offspring were mated to Prn-$p^{0/0}$ mice and the litters were screened for the presence of transgenes and absence of Prn-p^{+}-alleles by genomic Southern analysis or PCR. Since the PrP reading frame is intact, the PrP has the same amino acid sequence as the wild type, but it results in a slightly different expression pattern of PrP in the brain and possibly other expression sites, compared to wild type.[33] Thus for example, no PrP RNA was detected in Purkinje cells of *tg*a-20, but otherwise the mice showed high expression of PrP in the brain, as determined by immunoblots and *in situ* hybridization with a PrP antisense probe. These mice were highly susceptible to mouse prions.[15,34]

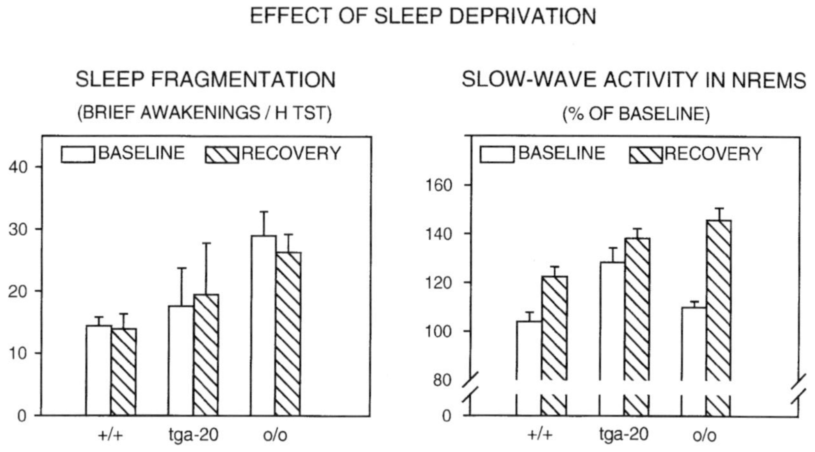

FIGURE 16.2
Effect of 6 h sleep deprivation on sleep fragmentation and slow-wave activity (mean EEG power density in the 0.5 to 4.0 Hz band) in nonREM sleep (NREMS) in the three genotypes: Prn-$p^{0/0}$ (o/o), Prn-$p^{+/+}$ (+/+), and transgenic mice overexpressing prion protein (*tga*-20). Bars represent mean values ± SEM, n = 8 for Prn-$p^{+/+}$ and Prn-$p^{0/0}$, n = 3 for *tga*-20. Left: Sleep fragmentation is expressed as the number of waking episodes <8 sec per hour of total sleep time (H TST) for the 12-h baseline period and 6 h recovery. Sleep fragmentation was significantly higher in the null-mice compared to the wild-type over the entire experiment ($p < 0.05$; *t*-test). Right: Bars represent slow-wave activity in NREMS for the first 2-h interval of baseline beginning after lights on and the first 2-h recovery interval each expressed as percentage of the 12-h baseline. Slow-wave activity was significantly increased in recovery in both genotypes, but the increase was significantly higher in the null-mice ($p < 0.05$; paired and unpaired *t*-test, respectively).

III. Results

Figure 16.2 illustrates the differences in sleep between the Prp-null mice, the wild type of the 129/Sv strain, and the transgenic *tg*a-20 mice overexpressing PrP. Sleep fragmentation was significantly larger, and the increase of EEG slow-wave activity in nonREM sleep was significantly higher in the null mice compared to the wild type. The results in the *tg*a-20 mice were either intermediate between the other two genotypes or more similar to the wild type, i.e., the reintroduction of PrP restored characteristics of the wild type in the null mice. In addition, both the outbred and inbred strains showed a significantly longer circadian activity period in constant darkness (DD) and a stability of period during prolonged DD compared to the wild type. Again, the *tg*a-20 mice were intermediate between the wild-type and null mice.[17]

IV. Outlook

Understanding the normal function of PrP may be of major importance in understanding the pathology in TSEs and in developing potential therapeutic approaches to the diseases. The involvement of PrP in the regulation of circadian rhythms has been validated by experimental testing of two independently generated lines of $PrP^{0/0}$ mice. Furthermore, the circadian rhythm phenotype was partially restored in transgenic mice overexpressing PrP, and preliminary results also indicate that sleep in the *tg*a-20 mice resembles that of the wild type more than that of the null mice. The experimental assessment of natural and artificial variants of PrP may enable the identification of structural elements in PrP involved in its natural function and possibly link them to elements involved in prion pathology.

It has now been shown that there is little or no phenotypic effect in many null mutant mice produced by gene targeting. This lack of phenotype is thought to be due to the organism compensating for the loss of a gene by alteration in expression of other genes or use of alternative developmental pathways.

Gene targeting has provided us with an extremely powerful approach for analyzing the prion protein diseases. The introduction of specific mutations will allow the role of these mutations in PrP to be defined. Gene targeting technology is now being developed through bacteriophage PI cre/lox site-specific recombinase system[35-37] to allow inducible gene expression, both tissue specifically and temporally. Inducible expression will allow the development of more specific models which ablate PrP gene expression at different times in development. These models will be able to differentiate between normal function of the PrP protein in the adult mouse and compensatory effects resulting from the complete ablation of PrP during development.

Transgenic models are now being produced by introducing specific alterations into the endogenous PrP gene using gene targeting techniques.[38] These will enable us to model FFI in mice and to examine the relationship between sleep, circadian rhythms, and PrP^{Sc} pathology.

References

1. Weissmann, C., Molecular biology of transmissible spongiform encephalopathies, *FEBS Lett.*, 389, 3, 1996.
2. Medori, R., Tritschler, H.-J., LeBlanc, A., Villare, F., Manetto, V., Chen, H. Y., Xue, R., Leal, S., Montagna, P., Cortelli, P., Tinuper, P., Avoni, P., Mochi, M., Baruzzi, A., Hauw, J. J., Ott, J., Lugaresi, E., Autilio-Gambetti, L., and Gambetti, P., Fatal familial insomnia, a prion disease with a mutation at Codon 178 of the prion protein gene, *N. Engl. J. Med.*, 326, 444, 1992.
3. Goldfarb, L. G., Haltia, M., Brown, P., Nieto, A., Kovanen, J., McCombie, W. R., Trapp, S., and Gadjusek, G. D., New mutation in scrapie amyloid precursor gene (at codon 178) in Finnish Creuzfeldt-Jakob kindred, *Lancet*, 337, 425, 1991.
4. Lugaresi, E., Medori, R., Montagna, P., Baruzzi, A., Cortelli, P., Lugaresi, A., Tinuper, P., Zucconi, M., and Gambetti, P., Fatal familial insomnia and dysautonomia with selective degeneration of thalamic nuclei, *N. Engl. J. Med.*, 315, 997, 1986.
5. Collinge, J., Palmer, M. S., Sidle, K. C., Gowland, I., Medori, R., Ironside, J., and Lantos, P., Transmission of fatal familial insomnia to laboratory animals, *Lancet*, 346, 569, 1995.

6. Hsiao, K. K., Scott, M., Foster, D., Groth, D. F., DeArmond, S. J., and Prusiner, S. B., Spontaneous neurodegeneration in transgenic mice with mutant prion protein, *Science*, 250, 1587, 1990.
7. Yu-Cheng, L. J., Lebo, R. V., Clawson, G. A., and Smuckler, E. A., Human prion protein cDNA: molecular cloning, chromosomal mapping, and biological implications, *Science*, 233, 364, 1986.
8. Sparkes, R. S., Simon, M., Cohn, V. H., Fournier, R. E. K., Lem, J., Klisak, I., Heinzmann, C., Blatt, C., Lucero, M., Mohandas, T., DeArmond, S. J., Westaway, D., Prusiner, S. B., and Weiner, L.P., Assignment of the human and mouse prion protein genes to homologous chromosomes, *Proc. Natl. Acad. Sci. USA*, 83, 7358, 1986.
9. Manson, J., McBride, P., and Hope, J., Expression of the PrP gene in the brain of Sinc congenic mice and its relationship to the development of scrapie. *Neurodegeneration,* 1, 45, 1992.
10. Kretzchmar, H. A., Prusiner, S. B., Stowring, L. E., and DeArmond, S. J., Scrapie prion proteins are synthesised in neurones, *Am. J. Pathol.*, 122, 1, 1986.
11. Moser, M., Colello, R. J., Pott, U., and Oesch, B., Developmental expression of the prion protein gene in glial cells, *Neuron*, 14, 509, 1995.
12. Caughey, B., Race, R., and Chesebro, B., Detection of prion protein mRNA in normal and scrapie-infected tissues and cell lines, *J. Gen. Virol.*, 69, 711, 1988.
13. Manson, J., West, J. D., Thomson, V., McBride, P., Kaufman, M. H., and Hope, J., The prion protein gene: a role in mouse embryogenesis? *Development*, 115, 117, 1992.
14. Büeler, H., Fischer, M., Lang, Y., Bluethmann, H., Lipp, H. P., DeArmond, S. J., Prusiner, S. B., Aguet, M., and Weissmann, C., Normal development and behaviour of mice lacking the neuronal cell-surface PrP protein, *Nature*, 356, 577, 1992.
15. Büeler, H., Aguzzi, A., Sailer, A., Greiner, R. A., Autenried, P., Aguet, M., and Weissmann, C., Mice devoid of PrP are resistant to scrapie, *Cell*, 73, 1339, 1993.
16. Manson, J. C., Clarke, A. R., Hooper, M. L., Aitchison, L., McConnell, I., and Hope, J., 129/Ola mice carrying a null mutation in PrP that abolishes mRNA production are developmentally normal, *Molec. Neurobiol.*, 8, 121, 1994.
17. Tobler, I., Gaus, S. E., Deboer, T., Achermann, P., Fischer, M., Rülicke, T., Moser, M., Oesch, B., McBride, P. A., and Manson, J. C., Altered circadian activity rhythms and sleep in mice devoid of prion protein, *Nature*, 380, 639, 1996.
18. Tobler, I., Deboer, T., and Fischer, M., Sleep and sleep regulation in normal and prion protein deficient mice, *J. Neurosci.*, 17, 1869, 1997.
19. Sforza, E., Montagna, P., Tinuper, P., Cortelli, P., Avoni, P., Ferrillo, F., Petersen, R., Gambetti, P., and Lugaresi, E., Sleep-wake cycle abnormalities in fatal familial insomnia. Evidence for the role of the thalamus in sleep regulation, *EEG Clin Neurophysiol.*, 94, 398, 1995.
20. Portaluppi, F., Cortelli, P., Avoni, P., Vergnani, L., Maltoni, P., Pavani, A., Sforza, E., Degli Uberti, E.C., Gambetti, P., and Lugaresi, E., Progressive disruption of the circadian rhythm of melatonin in fatal familial insomnia, *J. Clin. Endocrinol. Metab.*, 78, 1075, 1994.
21. Portaluppi, F., Cortelli, P., Avoni, P., Vergnani, L., Maltoni, P., Pavani, A., Sforza, E., Manfredini, R., Montagna, P., Roiter, I., Gambetti, P., Fersini, C., and Lugaresi E., Dissociated 24-hour patterns of somatotropin and prolactin in Fatal Familial Insomnia, *Neuroendocrinol.*, 61, 731, 1995.

22. Gambetti, P., Parchi, P., Petersen, R. B., Chen, S. G., and Lugaresi, E., Fatal familial insomnia and familial Creuzfeldt-Jakob disease: Clinical, pathological and molecular features, *Brain Pathology*, 5, 43, 1995.
23. Steriade, M., McCormick, D. A., and Sejnowski, T. J., Thalamocortical oscillations in the sleeping and aroused brain, *Science*, 262, 679, 1993.
24. Steriade, M., Contreras, D., and Amzica, F., Synchronized sleep oscillations and their paroxismal developments, *TINS*, 17, 199, 1994.
25. Moore, R. Y., and Speh, J. C., GABA is the principal neurotransmitter of the circadian system, *Neurosc. Lett.*, 150, 112, 1993.
26. Collinge, J., Whittington, M. A., Sidle, K. C. L., Smith, C. J., Palmer, M. S., Clarke, A. R., and Jefferys, J. G. R., Prion protein is necessary for normal synaptic function, *Nature*, 370, 295, 1994.
27. Lu, P., Sturman, J. A., and Bolton, D. C., Altered GABA distribution in hamster brain is an early molecular consequence of infection by scrapie prions, *Brain Res.*, 681, 235, 1995.
28. Clinton, J., Forsyth, C., Royston, M. C., and Roberts, G. W., Synaptic degeneration is the primary neuropathological feature in prion disease: a preliminary study, N*euroReport*, 4, 65, 1993.
29. Basler K., Oesch, B., Scott, M., Westaway, D., Wälchli, M., Groth, D. F., McKinley, M. P., Prusiner, S. B., and Weissmann, C., Scrapie and cellular PrP insoforms are encoded by the same chromosomal gene, *Cell*, 46, 417, 1986.
30. Liao, Y. C. J., Lebo, R. V., Clawson, G. A., and Smuckler, E. A., Human prion protein cDNA: molecular cloning, chromosomal mapping, and biological implications, *Science*, 233, 364, 1986.
31. Westaway, D., Cooper, C., Turner, S., Da Costa, M., Carlson, G. A., and Prusiner S. B., Structure and polymorphism of the mouse prion protein gene, *Proc. Natl. Acad. Sci. USA*, 91, 6418, 1994.
32. Gerlai, R., Gene-targeting studies of mammalian behavior: is it the mutation or the background genotype? *TINS*, 19, 177, 1996.
33. Fischer, M., Rülicke, T., Raeber, A., Sailer, A., Moser, M., Oesch, B., Brandner, S., Aguzzi, A., and Weissmann, C., Prion protein (PrP) with amino terminal deletions restoring susceptibility of PrP knockout mice to scrapie, *EMBO Journal*, 15, 1255, 1996.
34. Sailer, A., Bueler, H., Fischer, M., Aguzzi, A., and Weissmann, C., No propagation of prions in mice devoid of PrP, *Cell*, 77, 967, 1994.
35. Orban, P. C., Chui, D., and Marth, J. D., Tissue- and site-specific DNA recombination in transgenic mice, *Proc. Natl. Acad. Sci. USA*, 89, 6861, 1992.
36. Lasko, M., Sauer, B., Mosinger, B. Jr., Lee, E. J., Manning, R. W., Yu, S.-H., Mulder, K. L., and Westphal, H., Targeted oncogene activation by site-specific recombination in transgenic mice, *Proc. Natl. Acad. Sci. USA*, 89, 6232, 1992.
37. Gu, H., Marht, J., Orban, P., Mossman, H., and Rajewsky, K., Deletion of a DNA polymerase B gene segment in T cells using cell type specific gene targeting, *Science*, 265, 103, 1994.
38. Moore, R., Redhead, N., Selfridge, J., Hope, J., Manson, J., and Melton, D., Double replacement gene targeting for the production of a series of mouse strains with different prion protein gene alterations, *Biotechnology*, 13, 999, 1995.

Chapter 17

Measurement of Nitric Oxide in the Brain Using the Hemoglobin Trapping Technique Coupled with *In Vivo* Microdialysis

Julie A. Williams, Steven R. Vincent, and Peter B. Reiner

Contents

I. Introduction

Until recently, studies on nitric oxide (NO) physiology *in vivo* were limited to the use of NO synthase (NOS) inhibitors. This work has suggested a role for NO in

0-8493-3361-X/98/$0.00+$.50

modulation of neurotransmitter release,[1-5] regulation of local cerebral blood flow,[6-7] circadian rhythms,[8-9] behavioral state,[10-14] and arousal.[15-16] While these studies have provided valuable information with respect to potential roles of NO in the brain, they have not always been consistent or necessarily definitive. Moreover, interpretation of the results is severely compromised by the fact that NOS is widely distributed both in brain and body, resulting in innumerable and complex confounds.

An alternative approach is to study NO physiology on a regional basis. In order to accomplish this goal, we have used the hemoglobin-trapping technique combined with *in vivo* microdialysis to measure NO production across behavioral states. We have found that NO was produced in the thalamus at equal rates during wake and REM sleep, and at significantly lower rates during slow wave sleep (SWS).[17-18] This finding is consistent with our previous study which showed a similar profile for thalamic acetylcholine release.[19] Retrograde tracing combined with choline acetyltransferase immunohistochemistry demonstrated that the majority of the cholinergic inputs to the vicinity of the dialysis probe originated in the mesopontine tegmentum. It has been well established that these brainstem cholinergic neurons highly express NOS,[20-21] and provide the majority of NOS-containing inputs to the thalamus.[22] Electrical stimulation of the laterodorsal tegmental nucleus (LDT) significantly increased thalamic NO production, an effect which was blocked by a voltage-dependent sodium channel antagonist, tetrodotoxin.[18] Taken together, these data suggest that the state-dependent changes of thalamic NO production may have derived from the corresponding activity of mesopontine cholinergic/nitrergic neurons. Indeed, this is supported by extracellular recordings of LDT/PPT neurons *in vivo*, which showed that the majority of these cells exhibited the highest firing rate during EEG desynchronized states, wake and REM sleep.[23-24]

The hemoglobin trapping technique is not limited to measurement of thalamic NO production; we have also obtained preliminary results from measuring NO production in the hippocampus across behavioral states. The hippocampus contains NOS-positive interneurons[22] and receives substantial input from the cholinergic/NOS containing neurons in the medial septum,[25] a region which is hypothesized to play an important role in regulating cortical arousal.[26-27] The hemoglobin-trapping technique combined with *in vivo* microdialysis provides a relatively simple and inexpensive approach to assay NO production *in vivo* in freely moving animals.

II. Protocol

A. Assay of Nitric Oxide

Theory. As reviewed previously, the hemoglobin-trapping technique has been used to detect NO production in a variety of applications.[28-30] The technique is based upon the stoichiometric oxidation of the ferrous form of hemoglobin (oxyhemoglobin, or HbO_2) by NO[29] to produce the ferric form, metHb and nitrate. Others have shown previously a linear relationship between NO production and the conversion of HbO_2 to metHb by NO *in vitro*.[30-31] This technique has been used for both *in vitro* and *in vivo* applications.[32-34] The oxidation of HbO_2 can be quantified using absorbance spectrophotometry, because the color of HbO_2 changes from red to brown with oxidation. The Beer-Lambert Law describes the absorbance density (D) of a sample as proportional to its concentration (c) and optical pathlength (ℓ, defined as the length of the light pathway through the sample-containing cuvette):

$$D = \varepsilon \times c \times \ell \tag{1}$$

(for review, see Van Assendelft,[35] Feelisch et al.[28]). The molar extinction coefficient, ε, is a constant defined as the optical density of a sample at 1 mmol/l, measured with a pathlength of one cm. The same equation can also be applied to describe a *change* in concentration (i.e., $\Delta D = \Delta\varepsilon \times \Delta c \times \ell$), which is essentially what is being measured in the hemoglobin trapping technique: a decrease in the concentration of HbO_2 as it is converted to metHb.

Both HbO_2 and metHb exhibit absorbance maxima at several different optical wavelengths.[35] The key to selecting a wavelength for sample analysis is to select those at which the greatest difference occurs between HbO_2 and metHb. Several laboratories have used 401 nm, where metHb exhibits maximum absorbance. In this case, an *increase* in absorbance at 401 nm is directly related to the amount of metHb formed in an HbO_2 solution exposed to NO. For reasons described below, we measured absorbance changes at 577 nm, where HbO_2 exhibits an absorbance maximum.[35] Therefore, a *decrease* in the absorbance density at 577 nm is indicative of an increase in metHb formation. There are several different approaches to evaluating the absorbance changes in the samples depending upon the type of spectrophotometer available.[28] In some cases, the control HbO_2 solution may be used as the reference, and the absorbance change measured in a sample of equal volume that has been exposed to NO. In our situation, an *isosbestic* point (a wavelength where the absorbance densities of HbO_2 and metHb are equal) was used as an internal reference, since the absorbance at this wavelength does not change as HbO_2 is oxidized to metHb. It is best to use an isosbestic point that is closest to the test wavelength in order to minimize error due to nonspecific changes in background absorbance. If absorbance changes are measured at 401 nm, 410.5 nm should be used as the isosbestic; we used the isosbestic point at 591 nm.[35]

In order to quantify a change in HbO_2 concentration in samples which have been exposed to unknown concentrations of NO, it is necessary to construct a calibration curve to calculate an extinction coefficient. Our approach to measuring absorbance changes in samples involved obtaining two separate determinations of absorbance density differences between 577 and 591 nm — one for the control solution containing HbO_2, and one for the solution which had been exposed to NO. For the calibration, excess NO was added to HbO_2 solutions ranging in concentration from 0.5 to 5 μM, in order to ensure oxidation of the entire sample and thus produce a known concentration of metHb. The absorbance difference obtained for metHb was subtracted from that obtained from the corresponding concentration of HbO_2. The resulting absorbance values are plotted along the ordinate versus the hemoglobin concentration plotted along the abscissa. The slope of the resulting regression line is the extinction coefficient. Mathematically, this ε is described as:

$$\Delta\varepsilon_{577\text{-}591(\text{HbO2-metHb})} = \varepsilon_{577\text{-}591(\text{HbO2})} - \varepsilon_{577\text{-}591(\text{metHb})} \quad (2)$$

substituting equation (1) for $\varepsilon_{577\text{-}591(\text{HbO2})}$ and $\varepsilon_{577\text{-}591(\text{metHb})}$, $\Delta\varepsilon_{577\text{-}591(\text{HbO2-metHb})}$ can be obtained from known values:

$$\Delta\varepsilon_{577-591(\text{HbO2-metHb})} = \frac{\Delta D_{577-591(\text{HbO2})}}{[\text{HbO}_2]*\ell} - \frac{\Delta D_{577-591(\text{metHb})}}{[\text{metHb}]*\ell}$$

Perhaps in simpler terms, the extinction coeffient is as easy as y = mx + b. Since the absorbance differences are directly proportional to concentration changes, the y-intercept, b, should approximate zero; x would correspond to the concentration change, or amount of oxidation, which is the unknown; y is the measured absorbance difference between the control solution and the sample. As described above, the extinction coefficient is simply the slope of the line, m, which remains constant.

Preparation of Oxyhemoglobin. HbO_2 is prepared from bovine, double crystallized hemoglobin (Sigma) as follows. Prepare a sodium phosphate buffer solution, 50 m*M*, pH 7.4 (or a pH that is identical to what will be used for the dialysate). Using this buffer, prepare a 1 m*M* hemoglobin solution by dissolving 17 mg in 1 ml. Also prepare a 1 ml solution of 10 m*M* sodium dithionite, again using the buffer solution, and set aside. Bubble approximately 30 mls of the sodium phosphate buffer with oxygen (we use 95% O_2/5% CO_2) at room temperature, or on ice (colder solutions tend to favor oxygen binding to hemoglobin) for 30 minutes, until saturated. Equilibrate a Sephadex G-25 column (Pharmacia) with 25 ml of the oxygenated sodium phosphate buffer. When the last of the oxygenated buffer has eluted into the column (i.e., the meniscus is level with the top of the gel), add the sodium dithionite solution immediately. As soon as the dithionite elutes into the column, add the hemoglobin and stop the flow for one minute. This allows enough time for the dithionite to react with the hemoglobin to reduce it to deoxyhemoglobin. After the hemoglobin elutes into the column, add the remainder of the buffer. Once the hemoglobin is reduced by the dithionite, the formation of oxyhemoglobin can be monitored as the color changes to bright red as the hemoglobin passes through the column and reacts with the oxygen in the buffer.[36] Collect only the reddest part of the eluent, as the tail ends are likely to contain impurities. In order to protect HbO_2 from auto-oxidation and superoxide radicals, superoxide dismutase (500 units/ml) and catalase (2000 units/ml) may be added immediately to the eluent. This stock solution may be aliquotted and stored at –80°C for several months.

Because the hemoglobin is likely to have diluted during oxygenation, it is necessary to estimate the concentration of the stock solution. Dilute 10 µl of the HbO_2 stock in 3 ml of the sodium phosphate buffer (room temperature, *not* oxygenated) and measure the absorbance at 577 nm if the stock solution contains catalase. (Catalase exhibits a strong absorbance in the Soret region, which changes as it reacts with peroxide.[29] For this reason, our samples were analyzed using wavelengths of 577 and 591 nm.) Calculate the concentration of HbO_2 using an extinction coefficient of 15.37 mM^{-1} cm^{-1} (using a cuvette with a pathlength of 1 cm).[35] If the stock solution does not contain catalase, the absorbance may be measured at 415 nm and the concentration determined using an extinction coefficient of 131 mM^{-1} cm^{-1}. Usually, these diluted solutions approximate 2 μ*M*.

Calibration. Any spectrophotometer with the ability to measure absorbance densities of small-volume samples is acceptable for measurement of the conversion of HbO_2 to metHb; we use a Biorad diode-array spectrophotometer containing a flow-through cell and measure absorbance at 577 and 591 nm. The volume of the flow cell was 9 μl, with an optical pathlength of 6 mm. A Waters HPLC pump was used to deliver 100 μl samples at a rate of 1.0 ml per minute, and peak absorbance values were recorded for each wavelength.

For construction of the calibration curves, we prepared a solution containing 2 m*M* NO by bubbling NO gas in the sodium phosphate buffer (50 m*M*, pH 7.4), and adding this solution to known concentrations of HbO_2 (see below). NO-donor compounds may also be used, but the experimenter should be aware of other molecules these chemicals release which may also contribute to the oxidation or destruction of HbO_2, such as superoxide radicals or cyanide.

Preparation of aqueous NO takes place **in a fume hood** as follows. Place 20 ml of sodium phosphate buffer (or desired vehicle) in a vacuum filter flask (an Erlenmyer flask with a side vent). The top of the flask is sealed with a rubber stopper into which is inserted two PE-10 polyethylene tubes: one is attached to a tank of argon or helium and is used to remove oxygen from the phosphate buffer by bubbling it for 40 minutes *prior* to bubbling it with NO; the second tube is connected to the outlet from an NO reserve (see below). Attached to the side vent of the vacuum flask is a thick piece of tygon tubing (20 cm in length) with another piece of PE-10 tubing threaded inside. When bubbling the phosphate buffer, the end of this tubing acts as an exhaust. The PE-10 tubing is used to withdraw saturated NO solution from the flask using a gas-tight syringe. Therefore, there must be sufficient tubing inside the flask to reach to the bottom where the solution sits.

Prior to bubbling the solution, the outlet from the NO reserve is assembled as follows. The outlet tubing from an NO tank is attached (via rubber stopper) to a 20 cc glass syringe which contains KOH crystals and the tip is plugged with glass wool (the purpose of the KOH is to reduce any nitrogen oxide compounds back into the NO radical). A three-way stopcock is attached to the end of the syringe. The two outlets of this stopcock are connected to: 1) the exhaust, which consists of wide tubing aimed toward the *back* of the fume hood; and 2) fine (PE-10) tubing, which is passed through a sealed rubber stopper and into the vacuum flask (above) containing the buffer to be bubbled with NO. The three-way stopcock is used to flush oxygen out of the system: when the NO gas is turned on, the exhaust outlet from the stopcock should be open and the outlet to the flask should be closed. As the tubing from the NO reserve fills with NO, the gas inside appears brown, due the NO reacting with oxygen to form nitrogen dioxide and other similar compounds[37] which are very poisonous. When the tubing clears, you are now ready to bubble the solution in the vacuum flask.

Once the vacuum flask *and* NO reserve are oxygen-free, turn off the argon or helium, close the exhaust outlet from the glass syringe, and bubble the solution with NO for 10 minutes. A saturated NO solution at room temperature is approximately 2

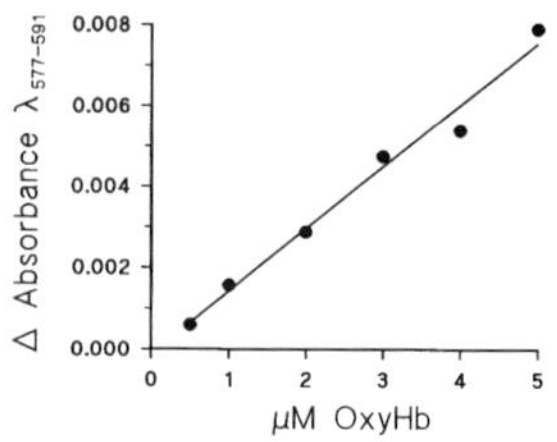

FIGURE 17.1
Calibration curve obtained by quantitative oxidation of increasing concentrations of HbO_2 to metHb (0.5 to 5 μ*M*) by aqueous NO. Values along the y-axis are the absorbance differences between 577 and 591 nm obtained for metHb subtracted from that obtained from corresponding concentrations of HbO_2. NO concentrations were estimated using the slope of the line. (From Williams, J. A., Vincent, S. R., and Reiner, P. B., Nitric oxide production in rat thalamus changes with behavioral state, local depolarization, and brainstem stimulation, *J. Neurosci.*, 17, 420, 1997. With permission.)

m*M*.[37] Prepare two sets of HbO_2 solutions in volumes of 2.5 ml which will range in concentration from 0.5 to 5 μ*M* after addition of vehicle or of aqueous NO to a final volume of 3 ml. Using a gas-tight syringe, extract the NO solution and transfer 0.5 ml each to one set of the HbO_2 solutions. Add 0.5 ml vehicle to each vial in the other set. Measure the absorbance differences, and plot on a graph as shown in Figure 17.1.

The calculated ε was 2.55×10^{-3} μM^{-1} cm^{-1}. Taking into account the 6 mm pathlength of the flow cell, the theoretical detection limit of the assay is 6.5 n*M* per 100 μl sample. We attempted to determine the detection limit empirically by adding known concentrations of NO to solutions containing 1 μ*M* HbO_2. The lower detection limit was 50 to 100 n*M*, a value which was considerably higher than the theoretical calculation. One possible explanation for this discrepancy is that a portion of the saturated NO solutions may have oxidized during handling for the sequential dilution. Thus, the actual concentrations of NO may have been reduced, and the lower detection limit over-estimated. The baseline concentrations of NO in the dialysis samples were typically in the range of 300 n*M* in awake, freely moving animals.

B. Microdialysis and Sample Collection Procedure

Surgery. Male Wistar rats weighing 285 to 320 g are anesthetized with 50 mg/kg pentobarbital i.p., and supplemented as needed. For EEG recordings, animals are chronically implanted with three screw electrodes (two for cortical EEG and one over the cerebellum as a reference) and depth electrodes for the recording of hippocampal theta (AP –3.1, DV –3.0, ML –2.4 from bregma[38]). Electrode pins are held in place with an Amphenol strip connector and fixed to the skull with dental acrylic.

Construction and implantation of transverse microdialysis probes have been described in detail elsewhere,[39] and only brief mention will be made here. Probes are made of a cellulose ester microdialysis membrane (i.d. 0.20, o.d. 0.21 mm, molecular weight cut off 10 KDa) with an active surface of 7 mm in the case of the thalamus, and 11 mm in the case of the hippocampus delimited by application of epoxy resin. A sharpened length of tungsten wire is threaded through the membrane for support, and a stainless steel cannula (22 gauge; 15 mm) attached to one end. To prevent the membrane from buckling during placement, the free end of the membrane is glued to the tungsten support. For implantation, the probe is secured horizontally to a micromanipulator on the stereotaxic apparatus, and gently inserted through the brain through holes drilled into the temporal bone at stereotaxic coordinates from bregma: AP –3.3, DV –5.8 to –5.6 for the thalamus, or at AP –4.3, DV –3.3 for the hippocampus.[38] The probe is advanced until it protrudes from the hole drilled on the opposite side, and the active surface centered using reference points marked directly on the probe. The glued end of the membrane is cut, the tungsten wire removed, and a second cannula attached to the free end of the membrane. Both cannulae are secured to the parietal bone with screws and dental acrylic and serve as the probe inlet and outlet.

Microdialysis. Microdialysis and sample collection procedures are as described.[18,19] Following surgery, rats are housed individually in 35×35×25 cm Plexiglas cages and are given food and water ad libitum. On the first day after surgery, each animal is moved into a secluded recording room and attached to a polygraph for several hours for adaptation. Experiments are performed on the second day after surgery during the light phase of a 12:12-h light:dark cycle (from 8:00 to 20:00) between 10:00 and 17:00. The samples are collected as follows. The dialysis probes are perfused with artificial cerebrospinal fluid (ACSF) containing (in m*M*) 147 NaCl, 3 KCl, 1.3 $CaCl_2$, 1.0 $MgCl_2$, 1.0 $NaPO_4$ buffer, pH 7.4, and 1 μ*M* HbO_2 for the assay of NO. The perfusion rate is 5 μl per minute, controlled by a syringe pump (Harvard Apparatus), and samples are collected in volumes of 100 μl, or 20-minute fractions for online experiments. For each sample, absorbance differences measured at 577 and 591 nm are compared against an equal volume of a control dialysate solution. In order to ensure the stability of HbO_2, the dialysate is replaced each hour (6 ml were prepared from stock each morning and stored on dry ice as 1-ml aliquots for use later in the day). Both the probe inlet and outlet consist of polyethylene tubing (800 cm × 0.28 mm), which contains an inner fused-silica tubing (100 μm i.d., 200 μm o.d.; Polymicro Technologies, Inc.) to maintain an air-tight environment. Prior to sample collection, animals are dialyzed for 60 minutes in order to allow the brain to equilibrate with the perfusion solution.

For the behavioral experiments, all dialysate samples are collected into ice-cold microcentrifuge tubes marked wake, SWS, or REM. Samples that are collected in the wake vials include periods in which the animal was clearly alert with a desynchronized EEG and engaged in some sort of waking behavior such as grooming, eating, drinking, or exploring. SWS vials include periods in which the animal was in a sleeping position, eyes closed, and the EEG synchronized for >30 sec at a time. REM vials include the sleep state in which the animal exhibits muscle twitches, EEG desychrony, and theta activity. All samples from ambiguous and transition states are discarded. Samples are collected during 5- to 6-hour recording sessions and are stored on dry ice for off-line analysis immediately following the session.

Microdialysis Membranes. Cellulose ester membranes (C-DAK 4000) are used for the dialysis probes for the following reasons. First, the molecular weight cutoff of the cellulose ester is 10 kDa, making it less likely for HbO_2, a 64 kDa protein, to exit the membrane. Second, the cellulose ester membrane is able to withstand the greater back pressure imposed by the addition of the fused silica tubing as the outlet from the probe. We found that Hospal membranes (polyacrylonitrite sodium methallyl sulfonate copolymer) expanded both *in vitro* and *in vivo* under such conditions, allowing much of the HbO_2 contained in the sample to leak out of the probe. As described above, a decrease in absorbance at 577 nm should be indicative of an oxidation reaction, rather than leakage of the HbO_2 from the solution.

III. Examples of Results Obtained

A. Pharmacological Properties of Thalamic NO Release

We carried out the bulk of our experiments in rat thalamus; the results obtained from hippocampus are preliminary. In order to determine whether the HbO_2 oxidation in the samples was due to NOS-related activity, rats were treated with 50 mg/kg of the NOS inhibitor, N^{ω}-nitro-L-arginine (N-ARG). N-ARG significantly reduced baseline concentrations of thalamic NO in awake, freely moving animals from 1.31 ± .07 to 0.86 ± .06 pmols/min (p <.001, n = 4).

Neuronal NOS and endothelial NOS are known to be Ca^{2+}-dependent, while the inducible NOS found in microglia is not. When a Ca^{2+}-free dialysate was applied either to the hippocampus (n = 2) or to the thalamus (n = 2), no change in baseline concentration was detected. However, when 10 mM BAPTA, a Ca^{2+} chelator, was added to this dialysate, baseline concentrations were significantly reduced from 1.09 ± .05 to 0.80 ± .07 pmols/min (p <.01, n = 3; Figure 17.2A). We next tested whether potassium-induced depolarization, which should activate voltage-dependent calcium channels and thereby increase NOS activity, affected NO output as measured by the hemoglobin trapping technique. A solution containing 30 mM potassium (in place of an equimolar amount of NaCl in order to maintain osmolarity) significantly increased NO output in urethane-anesthetized animals from a baseline of 0.72 ± .06 to 1.15 ± .11 pmols/min (p <.001, n = 4). Finally, to determine whether the potassium-induced increase was Ca^{2+}-dependent, 30 mM potassium was added to a 10 mM BAPTA/Ca^{2+}-free solution. The BAPTA/Ca^{2+}-free solution opposed the effect of high potassium (n = 3). Taken together, these data strongly suggest that NO output as measured by the hemoglobin-trapping technique was generated by a calcium-dependent isoform of NOS.

B. State-Dependence NO Release

In the thalamus, the mean NO concentration during wake was 1.34 ± .07 pmol/min; during SWS, 0.97 ± .03 pmol/min; and during REM, 1.42 ± .12 pmol/min. A similar profile was observed in the hippocampus with mean NO concentrations of 1.63 ± .08 pmol/min during wake, 1.14 ± .11 during SWS, and 1.71 ± .02 during REM

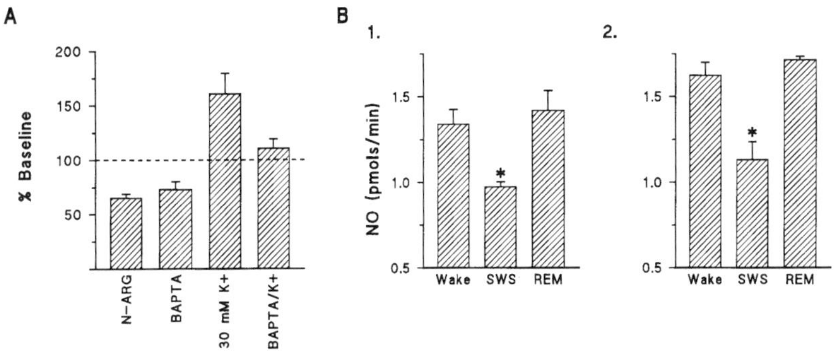

FIGURE 17.2
A) Pharmacology of thalamic NO production. Online experiments demonstrating that the HbO_2 oxidation measured in the samples was due to NO synthase and Ca^{2+}-dependent activity. NO concentrations are reported as percent baseline. Intraperitoneal injection of the NOS inhibitor, N-ARG, significantly reduced baseline concentrations in awake animals ($p < .001$; n = 4). Application of a Ca^{2+}-chelator, BAPTA (10 m*M*) in a Ca^{2+}-free dialysate solution significantly and reversibly reduced baseline concentrations in awake animals ($p < .01$; n = 3). Application of a solution containing 30 m*M* K^+ significantly increased NO release in urethane-anesthetized rats ($p < .001$; n = 4). Addition of 10 m*M* BAPTA to the high-K^+ solution blocked the potassium-induced increase (n = 3), suggesting that the increase was due to Ca^{2+}-dependent mechanisms. B) Extracellular NO concentrations across behavioral states in the rat thalamus; n = 7 (B1) and hippocampus; n = 3 (B2). Mean NO concentrations (± SEM) are reported for wake, slow-wave sleep (SWS), and REM sleep (n = 7). For both regions, NO concentrations did not differ between wake and REM, but were significantly lower during SWS ($p < .01$).

(Figure 17.2B). A repeated measures ANOVA showed that NO concentration varied significantly across states in both regions (thalamus: n = 7; hippocampus: n = 3; $p < .05$ with the Greenhouse-Geisser correction). Scheffe's *post hoc* analysis showed that NO concentration in both thalamus and hippocampus did not vary significantly between wake and REM but was significantly less during SWS ($p < .01$).

IV. Conclusion

The major advantage of the hemoglobin trapping technique coupled with *in vivo* microdialysis is that it permits one to correlate changes in the concentration of one of the smallest molecules in the body, NO, with one of its most macroscopic features, behavior, in freely moving animals. The pitfalls inherent in extrapolating between the two are many, and one should be appropriately prudent in deriving mechanistic implications from such data.

References

1. Guevara-Guzman, R., Emson, P. C., and Kendrick, K. M., Modulation of in vivo striatal transmitter release by nitric oxide and cyclic GMP, *J. Neurochem.*, 62, 807, 1994.
2. Strasser, A., McCarron, R. M., Ishii, H., Stanimirovic, D., and Spatz, M., L-arginine induces dopamine release from the striatum in vivo, *Neuroreport,* 5, 2298, 1994.

3. Silva, M. T., Rose, S., Hindmarsh, J. G., Aislaitner, G., Gorrod, J. W., Moore, P. K., Jenner, P., and Marsden, C. D., Increased striatal dopamine efflux in vivo following inhibition of cerebral nitric oxide synthase by the novel monosodium salt of 7-nitro indazole, *Br. J. Pharmacol.,* 114, 257, 1995.
4. Lonart, G., Wang, J., and Johnson, K. M., Nitric oxide induces neurotransmitter release from hippocampal slices, *Eur. J. Pharmacol.,* 220, 271, 1992.
5. Lorrain, D. S. and Hull, E. M., Nitric oxide increases dopamine and serotonin release in the medial preoptic area, *Neuroreport,* 5, 87, 1993.
6. Adachi, T., Inanami, O., and Sato, A., Nitric oxide (NO) is involved in increased cerebral cortical blood flow following stimulation of the nucleus basalis of Meynert in anesthetized rats, *Neurosci. Lett.,* 139, 201, 1992.
7. Northington, F. J., Matherne, G. P., and Berne, R. M., Competitive inhibition of nitric oxide synthase prevents the cortical hyperemia associated with peripheral nerve stimulation, *Proc. natl. Acad. Sci. USA,* 89, 6649, 1992.
8. Ding, J. M., Chen, D., Weber, E. T., Faiman, L. E., Rea, M. A., and Gilette, M. U., Resetting the biological clock: Mediation of nocturnal circadian shifts by glutamate and NO, *Science,* 266, 1713, 1994.
9. Weber, E. T., Gannon, R. L., Michel, A. M., gillette, M. U., and Rea, M. A., Nitric oxide synthase inhibitor blocks light-induced phase shifts of the circadian activity rhythm, but not c-*fos* expression in the suprachiasmatic nucleus of the Syrian hamster, *Brain Res.,* 692, 137, 1995.
10. Dzoljic, M. R. and De Vries, R., Nitric oxide synthase inhibition reduces wakefulness, *Neuropharmacol.,* 33, 1505, 1994.
11. Kapás, L., Fang, J., and Krueger, J. M., Inhibition of nitric oxide synthesis inhibits rat sleep, *Brain Res.,* 664, 189, 1994.
12. Dzoljic, M. R., De Vries, R., and van Leeuwen, R., Sleep and nitric oxide: effects of 7-nitro indazole, inhibitor of brain nitric oxide synthase, *Brain Res.,* 718, 145, 1996.
13. Leonard, T. O. and Lydic, R., Nitric oxide synthase inhibition decreases pontine acetylcholine release, *Neuroreport,* 6, 1525, 1995.
14. Leonard, T. O. and Lydic, R., Inhibition of nitric oxide synthase (NOS) in the medial pontine reticular formation (mPRF) decreases rapid eye movement (REM) sleep, *FASEB J.,* 10, A409, 1996.
15. Bagetta, G., Iannone, M., Del Duca, C., and nistico, G., Inhibition by N^{ω}-nitro-l-arginine methyl ester of the electrocortical arousal response in rats, *Br. J. Pharmacol.,* 108, 858, 1993.
16. Nistico, G., Bagetta, G., Iannone, M., and Del Duca, C., Evidence that nitric oxide is involved in the control of electrocortical arousal, *Ann. N.Y. Acad. Sci.,* 738, 191, 1994.
17. Williams, J. A., Vincent, S. R., and Reiner, P. B., NO escape from the thalamus and hippocampus, *Abstr. Soc. Neurosci.,* 1995.
18. Williams, J. A., Vincent, S. R., and Reiner, P. B., Nitric oxide production in rat thalamus changes with behavioral state, local depolarization, and brainstem stimulation, *J. Neurosci.,* 17, 420, 1997.
19. Williams, J. A., Comisarow, J., Day, J., Fibiger, H. C., and Reiner, P. B., State-dependent acetylcholine release in rat thalamus as measured by *in vivo* microdialysis, *J. Neurosci.,* 14, 5236, 1996.

20. Vincent, S. R., Satoh, K., Armstrong, D. M., and Fibiger, H. C., NADPH-diaphorase: A selective histochemical marker for the cholinergic neurons of the pontine reticular formation, *Neurosci. Lett.,* 43, 31, 1983.
21. Vincent, S. R., Satoh, K., Armstrong, D. M., Panula, P., Vale, W., and Fibiger, H. C., Neuropeptides and NADPH-diaphorase activity in the ascending cholinergic reticular system of the rat, *Neuroscience,* 17, 167, 1986.
22. Vincent, S. R. and Kimura, H., Histochemical mapping of nitric oxide synthase in the rat brain, *Neuroscience,* 46, 755, 1992.
23. El Mansari, M., Sakai, K., and Jouvet, M., Unitary characteristics of presumptive cholinergic tegmental neurons during the sleep-waking cycle in freely moving cats, *Exp. Brain Res.,* 76, 519, 1989.
24. Steriade, M., Datta, S., Pare, D., Oakson, G., and Curro Dossi, R., Neuronal activities in brain-stem cholinergic nuclei related to tonic activation processes in thalamocortical systems, *J. Neurosci.,* 10, 2541, 1990.
25. Kitchener, P. D. and Diamond, J., Distribution and colocalization of choline acetyltransferase immunoreactivity and NADPH diaphorase reactivity in neurons within the medial septum and diagonal band of Broca in the rat basal forebrain, *J. Comp. Neurol.,* 335, 1, 1993.
26. Buzsaki, G., Bickford, R. G., Ponomareff, G., Thal, L. J., Mandel, R., and Gage, F. H., Nucleus basalis and thalamic control of neocortical activity in the freely moving rat, *J. Neurosci.,* 8, 4007, 1988.
27. Semba, K., The cholinergic basal forebrain: a critical role in cortical arousal, in *The Basal Forebrain: Anatomy to Function,* Napier, T. C., Kalivas, P. W., and Hanin, I. (Eds.), Plenum, New York, 1991, 197.
28. Feelisch, M., Kubitzek, D., and Werringloer, J., The oxyhemoglobin assay, in *Methods in Nitric Oxide Research,* Feelisch, M. and Stamler, J. S. (Eds.), Wiley, Chichester, 1996.
29. Murphy, M. E. and Noack, E., Nitric oxide assay using hemoglobin method, *Methods in Enzymol.,* 233, 240, 1994.
30. Noack, E., Kubitzek, D., and Kojda, G., Spectrophotometric determination of nitric oxide using hemoglobin, *Neuroprotocols,* 1, 133, 1992.
31. Feelisch, M. and Noack, E., Correlation between nitric oxide formation during degradation of organic nitrates and activation of guanylate cyclase, *Eur. J. Pharmacol.,* 139, 19, 1987.
32. Murphy, M. E., Piper, H. M., Watanabe, H., and Sies, H., Nitric oxide production by cultured aortic endothelial cells in response to thiol depletion and replenishment, *J. Biol. Chem.,* 266, 19378, 1991.
33. Mayer, B., Klatt, P., Böhme, E., and Schmidt, K., Regulation of neuronal nitric oxide and cyclic GMP formation by Ca^{2+}, *J. Neurochem.,* 59, 2024, 1992.
34. Balcioglu, A. and Maher, T. J., Determination of kainic acid-induced release of nitric oxide using a novel hemoglobin trapping technique with microdialysis, *J. Neurochem.,* 61, 2311, 1993.
35. Van Assendelft, O. W., *Spectrophotometry of Haemoglobin Derivatives,* Van Gorcum, Groningen, 1970.

36. Dixon, H. B. F. and McIntosh, R., Reduction of methaemoglobin in haemoglobin samples using gel filtration for continuous removal of reaction products, *Nature,* 213, 399, 1967.
37. Feelisch, M., The biochemical pathways of nitric oxide formation from nitrovasodilators: Appropriate choice of exogenous NO donors and aspects of preparation and handling of aqueous NO solutions, *J. Cardiovasc. Pharmacol.,* 17 (Suppl. 3), S25, 1991.
38. Paxinos, G. and Watson, C., *The Rat Brain in Stereotaxic Coordinates,* Academic Press, New York, 1982.
39. Damsma, G. and Westerink, B. H. C., A microdialysis and automated on-line analysis approach to study central cholinergic transmission in vivo, in *Microdialysis in the Neurosciences,* Robinson, T. E. and Justice, J. (Eds.), Elsevier, Amsterdam, 1991, 237.

Chapter 18

Mapping Regional Cerebral Protein Synthesis During Sleep

Rebecca K. Zoltoski

Contents

I. Introduction

Sleep must serve an important physiological function. One possibility is summarized by the restorative hypothesis, which suggests that energy conservation during

0-8493-3361-X/98/$0.00+$.50

sleep favors the anabolic restoration of tissue components (notably of the central nervous system) that are supposedly deteriorated during wakefulness (for review see Adam, 1980).[1] The integrity of the cerebral tissue is restored by the synthesis of new macromolecules, in particular, new proteins. Proteins play a fundamental role in cellular structure and function. However, the results of the few studies of the effect of sleep on protein synthesis are controversial (for review see Maquet, 1995).[2] Briefly, in 1957, Shapot[3] reported that the incorporation of [^{35}S]methionine was increased in sleep rats in comparison with exhausted animals with phenamine and with control animals. Using similar techniques, Bobillier and colleagues[3a] reported that in cats receiving p-chlorophenylalanine and 5-hydroxytryptophan, there was a significant positive correlation between the duration of rapid eye movement sleep (REMS) as well as slow wave sleep (SWS) and the incorporation of amino acids within the proteins of the telencephalon. In a confounding study, Brodskii and colleagues[3b] measured the ability of cortical biopsies of cats to incorporate [^{3}H]leucine into proteins. After 20 min of non-REMS (NREMS) the incorporation of leucine was significantly lower than following 1.5 min of REMS or 20 min of wakefulness.

Autoradiographic studies using [^{3}H]leucine analyzing cerebral protein synthesis (CPS_{leu})[4-6] during sleep have also offered controversial results. Nakanishi and colleagues[7] reported that no modification of CPS_{leu} was noted between sleeping and awake monkeys. However, the authors combined both light and deep SWS into the sleeping category, which may have confounded their results. Alternatively, Ramm and Smith[8] reported a positive correlation between CPS_{leu} and percent weighted time spent in SWS following 28 to 72 hours of REMS deprivation, while no significant correlation was noted between CPS_{leu} and percent weighted time in REMS or wakefulness. Most recently, Smith and her colleagues[9] have presented preliminary data that seem to support this observation, such that in monkeys, a positive correlation exists between percent weighted time spent in deep sleep and CPS_{leu} in 57 regions of the brain studied.

Our goal is to further test the hypothesis that restoration does occur during SWS, by mapping rates of CPS using tracer autoradiography with L-[1-^{14}C]leucine during sleep and recovery sleep, following various periods of sleep deprivation. A single pulse injection of L-[1-^{14}C]leucine is given intravenously, arterial blood samples are collected for 45 min (while the animal is awake or sleeps), and the brain is then processed for autoradiography with a formalin fixation and washing of the tissue sections. The labeled leucine is taken up by the brain and either incorporated into protein or metabolized.[10] The resulting labeled CO_2 from the metabolic pathway is quickly removed from the brain by the circulation. Therefore, following the formalin fixation to remove free labeled leucine, the radioactivity measured from the autoradiograms reflects regional rates of amino acid incorporation into proteins during the sleep period.

II. Protocol

A. Surgical Procedure

1. Under deep anesthesia, insert catheters (PE-50 tubing) into the femoral vein and artery of one leg. Insert each catheter approximately 8 cm into each vessel. Secure to the abdominal muscle and thread these lines subcutaneously to exit the scalp at the nape of the neck.
2. Implant standard sleep recording electrodes. These can include a bipolar electrode into the hippocampus (AP –4.2, ML –3.4, DV –3.5 from bregma using the flat skull technique),[11] screw electrodes over the frontal and parietal cortices, and flexible wire electrodes into the neck muscles.
3. Secure all electrodes into a pedestal and secure the catheters to the pedestal using dental acrylic applied around a straw to allow easy access to the catheters. They can then be secured inside the straw with a plug.
4. Allow a three-day recovery in which the arterial catheter is flushed with heparinized saline (50 usp/cc) daily and the animal is checked for signs of infection around the incisions.

B. Experimental Procedure

1. Setup

1. Position the recording chamber such that you have access, but so that the animal is not disturbed during the injection and sampling procedure.
2. Provide white noise (70 dB) and heat (30°C).
3. Food deprive the animal for a minimum of three hours to ensure stabilization of plasma values.
4. Assess the initial physiological condition of the animal by recording rectal temperature and weight. Blood samples for hematocrit and glucose determination will be collected throughout the experimental procedure to further assess the physiological condition of the animal.
5. Connect the arterial and venous catheter to tubing that permits access from outside the cage.
6. Initiate polygraphic recording of the EEG and EMG to record the state of sleep throughout the experimental period.

2. Leucine Injection Protocol

1. Vary the behavioral state of the animal, such that equal numbers of animals are injected in each state.
2. Collect control blood samples from the arterial catheter once the animal has obtained the desired behavioral state. Collect 500 μl of blood in an Eppendorf tube. Fill a hematocrit tube and seal. Place both of these on ice until the completion of the study.

TABLE 18.1
Time of Collected Arterial Blood Samples During the Leucine Incorporation Protocol

Time of arterial blood sample	Volume collected	Analysis*
0	500 µl	LSC, AA, Glucose, HCT
0.5	250 µl	LSC, Glucose
1	250 µl	LSC, Glucose
2	250 µl	LSC, Glucose
3	250 µl	LSC, Glucose
5	250 µl	LSC, Glucose
10	500 µl	LSC, AA, Glucose, HCT
15	250 µl	LSC, Glucose, HCT
20	500 µl	LSC, AA, Glucose, HCT
30	500 µl	LSC, AA, Glucose, HCT
45	500 µl	LSC, AA, Glucose, HCT

* Abbreviations used: LSC = Liquid scintillation counts to determine content of L-[1-^{14}C]leucine; AA = Amino acid determination; and HCT = hematocrit.

3. Inject 125 µCi/kg L-[1-^{14}C]leucine as an intravenous pulse. This dose should be dissolved in physiological saline to a total volume of 0.5 to 0.7 ml. Follow the injection with a saline flush to ensure that all of the isotope has entered the blood stream. Start timing for sample collection when the leucine has been entirely injected into the tubing, just prior to the flush.
4. Collect timed arterial blood samples as shown in Table 18.1. Store on ice until the completion of the study. For the first 5 min of sampling, a steady slow flow of blood can be maintained. For the remainder, flush tubing with only enough heparinized saline to clear the tubing of blood so that no clots occur.
5. At the end of the 45-min incubation period, administer an overdose of pentobarbital (120 mg/kg) through the arterial catheter.
6. Rapidly extract the brain and freeze it in isopentane cooled to –40°C with dry ice. Store at –80°C until cutting.

3. *Blood Sample Processing*

1. Centrifuge the blood samples to separate the plasma from the red blood cells.
2. Determine the radioactive leucine content of each sample by first deproteinizing 50 µl of plasma with 50 µl of 10% perchloric acid or 5-sulfosalicylic acid. Centrifuge to pellet the protein precipitate and collect 50 µl of the supernatant into scinitillation vials containing 7.5 ml of counting cocktail for liquid scintillation counting.
3. Determine the free leucine content of each sample by diluting 20 µl of plasma with 280 µl of distilled water and then deproteinizing with 100 µl of 16% sulfosalicylic acid containing 0.04 m*M* norleucine as an internal standard. Vortex and store frozen

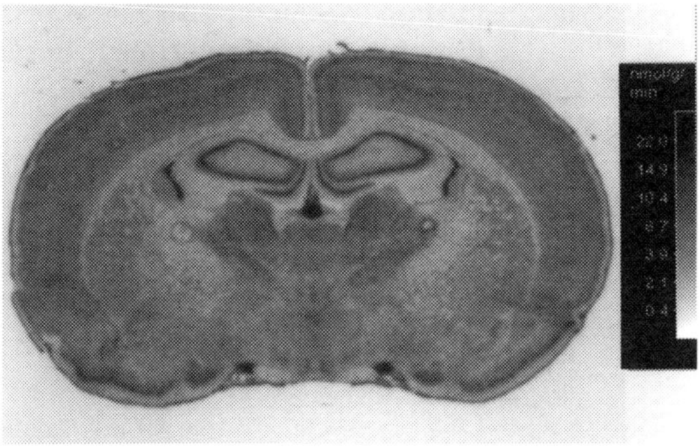

FIGURE 18.1
Representative section of a rat brain following the standard leucine injection protocol. (Image from MCID™ software-sample image file, Imaging Research, Inc., St. Catharines, Onario.)

at –80°C until analyzed for free amino acid content. It is important that the samples are deproteinized and frozen immediately after collection, because fluctuations in temperatures can cause a breakdown of proteins which will artificially elevate the free amino acid concentrations.[12]

4. Assess the physiological condition of the animal during the experimental period by determining glucose content of remaining plasma and the hematocrits.

4. *Tissue Processing*

1. Cut 20 μm sections in a cryostat at –20°C, thaw-mount onto gelatin-coated slides, and air dry.
2. Fix and wash with five changes of phosphate-buffered 10% formalin for 30 min each and wash in running deionized water for 30 min.
3. Expose the slides for 55 to 70 days on coated Kodak SB5 X-ray film along with [^{14}C]polymethylacrylate standardization scales (Amersham) for autoradiography.
4. Develop the film using standard procedures and dry thoroughly.
5. A representative autoradiogram is shown in Figure 18.1.

5. *Data Analysis*

1. The rates of leucine incorporation into protein in individual brain regions and the average of the brain as a whole can be determined by analysis of the autoradiograms and of the time course of [^{14}C]leucine and leucine in the plasma. The concentration of ^{14}C is measured from the autoradiogram and compared to the ^{14}C concentration curve derived for each film from the optical densities of the calibrated standards. Local rates of protein synthesis are calculated by using the following formula:[6]

$$R_i = \frac{P_i^*(T)}{\lambda_i \int_0^T [C_P^*(t)/C_P]\,dt} \qquad (1)$$

2. where R_i is the rate of leucine incorporation into protein in tissue i; $P_i{}^*(T)$ is the concentration of ^{14}C fixed in the tissue i at any time, T, after introduction of the tracer into the circulation; λ_i is an indication of the fraction of leucine in the precursor pool which is provided by the plasma; $C_p{}^*(t)$ and C_p are the concentration of labeled and unlabeled leucine in the arterial plasma, respectively; and t is variable time. The factor λ_i, a constant between 0 and 1.0, is the only variable not measured directly in each experiment, but is evaluated separately. In conscious, adult male rat brain as a whole, λ_i has been determined to be 0.58.[13]
3. This calculation is simplified by using a computerized image processing system.
4. Score the sleep record in 20-sec epochs.
5. Each epoch occurring during the leucine incubation period will be weighted by the tissue level of free L-[1-^{14}C]leucine available during the epoch as calculated using the following formula:[8]

$$\text{weighted time in state} = \frac{\sum_{i=n} \int_{si}^{ti} \text{Ce}^*(t)\,dt}{\int_0^T \text{Ce}^*(t)\,dt} \qquad (2)$$

6. where Ce*(t) is the tissue concentration of free L-[1-^{14}C]leucine in tissue at time t; i is the epoch of the state; si and ti indicate the length of the episode of the state; n is the total number of epochs of each state; and T is total time of the leucine incubation period. This formula reflects the importance of timing each epoch, in that the epochs occurring at the beginning of the leucine incubation period are more important than those occurring at the end. This procedure will weight each epoch by its relative importance when determining the amount of labeled leucine incorporated into proteins during the various sleep states.
7. Correlations between weighted time in a state and CPS_{leu} are used for statistical analysis.

III. Conclusions

A. Limitations of Techniques

The major source of error in L-[1-^{14}C]leucine method comes from the assumption that unlabeled leucine derived from protein degradation does not recycle into the precursor pool for protein synthesis. However, such recycling has been shown to occur[13] and is accounted for by factor λ_i in the operational equation of the method. The value of λ_i for conscious, adult, male rat brain as a whole (0.58) indicates that there is significant recycling (42%) of unlabeled leucine derived from the naturally occurring breakdown of proteins in brain. This value is a property of the tissue and the biological condition. Whether or not it is the same in various regions of the brain or under different physiological or pathological conditions, such as recovery sleep following deprivation, has not been determined. Using this value under other conditions may result in an error in estimation of actual CPS_{leu} levels, due to its estimated variability in gray matter of approximately 10%.[14] However, since only animals that exhibit normal body temperature, plasma leucine and glucose, hematocrit, and electrophysiological and behavioral characteristics are used in most sleep studies, this value for λ_i can be assumed to be accurate.

Another problem that is encountered in these studies is that the leucine incubation period contains a continuum of behavioral states. As most of the labeled leucine is incorporated into tissue during the first few minutes, those states which are strongest in the first 15 min of the incubation period will carry more relative importance. This is adequate for studying wakefulness and SWS; however, REMS is largely precluded from this type of analysis. However, use of the weighted values of behavioral states and correlative statistics will minimize this impact. Also, use of deprivation periods to analyze recovery sleep may play an important role in assessing the effect of REMS on CPS_{leu}.

B. Future Directions

Various methods of sleep deprivation have been used to analyze the function of sleep. However, the method of sleep deprivation may be a serious potential confound for studies of the metabolic and physiological consequences of elevated sleep need.[14,15] We have established a paradigm in which relatively short periods (up to about 48 hours) of sleep deprivation can be obtained with minimal stress. We will use this paradigm to characterize the effects of various periods of total and state-specific sleep deprivation upon rates of CPS_{leu}. This type of analysis will hopefully allow for a more direct test of the restorative hypothesis of sleep as well as to further understanding of the consequences of sleep deprivation.

References

1. Adam, K., Sleep as a restorative process and a theory to explain why, in *Adaptive Capabilities of the Nervous System, Progess in Brain Research*, 53rd Ed., McConnell, P. S., Boer, G. J., Romijn, H. J., van de Poll, N. E. and Corner, M. A. (Eds.), Elsevier, Amsterdam, 1980, 289.
2. Maquet, P., Sleep function(s) and cerebral metabolism, *Behav. Brain Res.*, 69, 75, 1995.
3. Shapot, V. S., Brain metabolism in relation to the functional state of the central nervous system, in *Metabolism of the Nervous System*, Richter, D. (Ed.), Elsevier/Pergamon Press, London, 1957, 247.

3a. Bobillier, P., Froment, J. L., Seguin, S., and Jouvet, M., Effets de la p-chlorophenylalanine et du 5-hydroxytrytophane sur le sommeil et le métabolisme central des monoamines et des protéines chez le chat, *Biochem. Pharmacol.,* 22, 3077, 1973.

3b. Brodskii, Vy., Gusatinskii, V. N., Kogan, A. B., and Nechaeva, N. V., Modifications of incorporation of [3H] leucine during slow wave sleep and REM sleep in the cerebral cortex of the cat, *Dokl. Akad. Nauk. SSSR,* 215, 748, 1974.

4. Ingvar, M. C., Maeder, P., Sokoloff, L., and Smith, C. B., Effects of ageing on local rates of cerebral protein synthesis in Sprague-Dawley rats, *Brain*, 108, 155, 1985.

5. Smith, C. B., Davidsen, L., Deibler, G. E., Patlak, C., Pettigrew, K., and Sokoloff, L., A method for the determination of local rates of protein synthesis in the brain, *Trans. Am. Soc. Neurochem.*, 11, 94, 1980.
6. Smith, C. B., The measurement of regional rates of cerebral protein synthesis in vivo, *Neurochem. Res*, 16(9), 1037, 1991.
7. Nakanishi, H., Kennedy, C., Smith, C. B., Sun, Y., Dang, T., and Sokoloff, L., Local rates of cerebral protein synthesis in non-REM sleep, *J. Cereb. Blood Flow Metab.*, 9 (suppl. 1), S737, 1989.
8. Ramm, P. and Smith, C. T., Rates of cerebral protein synthesis are linked to slow wave sleep in the rat, *Physiol. Behav.*, 48, 749, 1990.
9. Smith, C. B., Dang, T., Ito, M., Mori, K., Nakanishi, H., Nakamura, R. K., Namba, H., Storch, F., Suda, S., Sun, Y., Gillin, J. C., Mendelson, W., Mishkin, M., Kennedy, C., and Sokoloff, L., Local rates of cerebral protein synthesis measured in sleeping monkeys are positively correlated with percent time in deep sleep, *Soc. Neurosci. Abstr.*, 21, 1495, 1995.
10. Banker, G. and Cotman, C. W., Characteristics of different amino acids as protein precursors in mouse brain: advantages of certain carboxyl-labeled amino acids, *Arch. Biochem. Biophys.*, 142, 565, 1971.
11. Paxinos, G. and Watson, C., *The Rat Brain in Stereotaxic Coordinates*, Academic Press, Sidney, Australia, 1986.
12. Schaefer, A., Piquard, F., and Haberey, P., Plasma amino-acids analysis: effects of delayed samples preparation and of storage, *Clinica Chimica Acta*, 164, 163, 1987.
13. Smith, C. B., Deibler, G. E., Eng, N., Schmidt, K., and Sokoloff, L., Measurement of local cerebral protein synthesis in vivo: Influence of recycling of amino acids derived from protein degradation, *Proc. Natl. Acad. Sci USA*, 85, 9341, 1988.
14. Zoltoski, R. K. and Ramm, P., Effects of sleep deprivation on plasma free amino acids in the rat, *Soc. Neurosci. Abstr.*, 21, 953, 1995.
15. Ramm, P. and Zoltoski, R. K., Various types of sleep deprivation have different effects on cerebral protein synthesis in the rat, *Soc. Neurosci. Abstr.*, 21, 1679, 1995.

Index

A

B

C

E

F

G

H

I

L

M

N

O

P

R

S